Unclassified and Secure

A Defense Industrial Base Cyber Protection Program for Unclassified Defense Networks

DANIEL GONZALES, SARAH HARTING, MARY KATE ADGIE,
JULIA BRACKUP, LINDSEY POLLEY, KARLYN D. STANLEY

Approved for public release; distribution unlimited

For more information on this publication, visit **www.rand.org/t/RR4227**

Library of Congress Cataloging-in-Publication Data is available for this publication.
ISBN: 978-1-9774-0448-0

Published by the RAND Corporation, Santa Monica, Calif.

Preface

The defense industrial base (DIB) is under attack. Foreign actors are stealing large amounts of sensitive data, trade secrets, and intellectual property every day from DIB firms—contributing to the erosion of the DIB and potentially harming U.S. military capabilities and future U.S. military operations. In 2018, the U.S. Secretary of the Navy noted, "attacks on our networks are not new, but attempts to steal critical information are increasing in both severity and sophistication."[1] The U.S. Department of Defense (DoD) has taken steps to better secure systems against cyber threats, but most well-established protections in place focus on classified networks, while unclassified networks have become an attractive "backdoor" entrance for adversaries seeking access to cutting-edge technologies and research and development efforts. DoD simply lacks a comprehensive strategy for protecting the unclassified networks of DIB firms. To address this problem, DoD has increased regulations and introduced new security controls, but the current approach may be insufficient—DIB firms cannot keep up. Cybersecurity is necessary but also expensive—a suite of cybersecurity tools requires expertise to use, and the required combination of tools and skilled professionals may not be affordable for many DIB firms. Furthermore, the regulatory environment is complex and challenging to navigate, even for large firms with robust cybersecurity teams.

This report offers DoD a way ahead to better secure unclassified networks holding defense information through the establishment and implementation of a cybersecurity program designed to strengthen the protections of these networks. The program offers a means for DoD to better monitor the real-time health of the DIB and ensure that protections are in place to prevent the disclosure of sensitive corporate information from DIB firms or sensitive supply chain information across the DIB. The program also includes a means to offer qualified small DIB firms access to cybersecurity tools for use on unclassified networks, for free or at a discounted rate, to ensure that affordable protections are accessible to all DIB firms. To be sure, advanced persistent threats and sophisticated cyber attacks will not stop. However, this program can help build

[1] Gordon Lubold and Dustin Volz, "Chinese Hackers Breach U.S. Navy Contractors," *Wall Street Journal*, December 14, 2018.

stronger walls, develop more-coordinated responses, and reinforce the foundation on which so much of U.S. military power is built.

This report should be of interest to senior DoD decisionmakers, DIB stakeholders, and congressional leaders exploring options for protecting unclassified DIB networks used by small DIB firms with limited in-house cybersecurity expertise and tools.

Funding for this venture was made possible by the independent research and development provisions of RAND's contracts for the operation of its U.S. Department of Defense federally funded research and development centers.

Contents

Figures

Tables

Summary

Cyber attacks against defense industrial base (DIB) firms designed to steal sensitive data, trade secrets, and intellectual property (IP) and are growing in sophistication and severity. Cyber attacks designed to steal IP from the unclassified networks of U.S. companies have increased, with small firms particularly vulnerable given their challenges in affording the costly cybersecurity tools (CSTs) and skilled professionals required to adequately protect their networks. In addition, ransomware attacks, possibly by different perpetrators, have also recently increased and have resulted in the destruction of data held on the unclassified networks of small companies and local governments. Examples include the LockerGoga ransomware attack in March 2019 and the NotPetya attack in June 2017.[2]

Meanwhile, the current U.S. Department of Defense (DoD) approach to prevent these attacks from being successful is based on Defense Acquisition Regulation Supplement (DFARS) 252.204-7012 and National Institute of Standards and Technology (NIST) Special Publication (SP) 800-171 and appears to be inadequate. As of July 2019, no DIB firms have been able to fully implement the cybersecurity controls specified in NIST SP 800-171.[3] In addition, according to our cost analysis, small DIB firms and some medium-sized firms will not have the resources to comply with NIST SP 800-171. Furthermore, DFARS 252.204-7012 assumes that controlled unclassified information (CUI)—information deemed by DoD to require additional safeguards—flows down from the prime contractors, with primes responsible for denying a subcontractor access to CUI if the subcontractor does not comply with the regulation. However, many subcontractors are in business because of their trade secrets. CUI exists at all levels of the supply chain, and this information must be protected but is overlooked in the current DFARS clause.

In this report, we argue that the current approach for protecting a significant amount of CUI on DIB firm unclassified networks from cyber attacks conducted by

---

[2] Multi-State Information Sharing and Analysis Center, "LockerGoga," security primer, SP2019-0611, March 2019; U.S. Department of Homeland Security, "Alert (TA17-181A), Petya Ransomware," revised February 15, 2018.

[3] Sera-Brynn, *Reality Check: Defense Industry's Implementation of NIST SP 800-171: Keen Insights from Certified Cybersecurity Assessors*, May 2019.

foreign actors may be insufficient. The persistent attacks and hemorrhaging of critical information and technology from unclassified networks, coupled with associated significant financial losses, erodes the U.S. DIB and threatens U.S. military advantage over the long term. In this report, we offer an approach—a DIB Cyber Protection Program (DCP2)—for bolstering DoD protections for unclassified DIB firm networks and better positioning DIB firms to defend against this serious threat.

Findings

In this report, we define the DIB; estimate its composition of small, medium, and large DIB firms; and identify DIB populations most vulnerable to cyber intrusion and most disadvantaged in terms of being able to afford adequate protections. We discuss the current DoD approach for protecting unclassified networks from cyber attacks and assess the cost of cybersecurity—how much firms spend relative to what they should be spending—and the current cybersecurity landscape of cyber protection tools available.

Unclassified Networks of Small Defense Industrial Base Firms Are at Higher Risk

Our research reveals that the cybersecurity architectures of small DIB firms are likely to be deficient in several key areas: user authentication, network defenses, vulnerability scanning, software patching, and security information and event management (SIEM), or cyber attack response. Small DIB firms also probably lack other important CSTs that have been developed to respond to new threats, because any small firms cannot afford to procure and operate these CSTs. It is also important to highlight why SIEM systems are so important: Employing a purely perimeter defense-based cybersecurity architecture is unlikely to be successful. A SIEM capability is essential detect, isolate, and extract malware after it has gained access to the network.

Current DoD Approach Likely Unaffordable for Many Small and Some Medium-Sized Defense Industrial Base Firms

In our research, we found that the current DoD approach of policies and procedures for protecting CUI is unaffordable or inaccessible for key members of the DIB. It may not be feasible for small DIB firms to comply with the security control guidance issued by DoD and, as a result, may deter small firms from bidding on DoD contracts.

DoD's CMMC Process Is Likely Unaffordable for Small and Some Medium-Sized Defense Industrial Base Firms

One recent addition to the DoD approach is a compliance-based cyber maturity model and a mandatory certification process—the Cybersecurity Maturity Model Certification (CMMC). Our cost analysis indicates that most small DIB firms may not be able to afford the cyber defenses that could be mandated by the CMMC, and many medium-sized DIB firms may face the same challenges, especially if held to the highest compliance levels of the CMMC.

Current Voluntary DoD Cyber Threat Sharing Service Cannot Reach Many Defense Industrial Base Firms

The voluntary program for cyber threat information sharing poses challenges, as well, as not all DIB firms can access this service because it requires a DoD Common Access Card (CAC), which not all DIB firms or employees have. Some DIB firms may lack the informal ties to the Intelligence Community that would make them privy to important cyber threat information.

Advanced Cybersecurity Tools Can Strengthen Defense Industrial Base Cyber Defenses but Are Costly

Cybersecurity firms have developed advanced CSTs that would help strengthen the cybersecurity of the DIB, but these new tools are costly. In this report, we discuss many of these tools and how they could be leveraged by DIB firms and as part of the DCP2.

Recommendation

To address the cybersecurity vulnerabilities of the unclassified networks of DIB firms, we recommend DoD establish a DCP2 designed to

- **improve the monitoring and real-time health** of the DIB
- **improve cybersecurity** for firms that cannot afford the needed CSTs and professional staff
- **offer data protections** to prevent the disclosure of sensitive corporate information from DIB firms to DoD, prevent sensitive supply chain information from being disclosed across the DIB, and prevent the exfiltration of DIB sensitive data to adversaries
- **offer legal protections** for DIB firms, to minimize the liability that DIB firms might have if the information they provide to the government is used in unanticipated ways.

How the DIB Cyber Protection Program Would Work

The DCP2 would be a voluntary program under which DoD would provide CSTs to DIB firms either free of charge or at significantly reduced licensing costs. In turn, the DIB firms would agree to provide sanitized data produced by the CSTs to a security operations center (SOC)—either one run by DoD (DIB SOC) or a trusted third-party

SOC—devoted exclusively to defending the DIB.[4] These sanitized data would include network metadata, application metadata, anonymized user account metadata, security alerts, and anonymized system log files; they would not include the personally identifiable information of DIB firm employees, proprietary firm information, employee correspondence, or any CUI. DoD would provide, free of charge, a data-sanitization application to ensure that only relevant cybersecurity data are transmitted to the DIB SOC or commercial SOC.

The DIB SOC or commercial SOC would provide dynamic intelligence, security alerts, and recommended actions to DIB firms to identify and remediate advanced persistent threat incursions and to prevent the exfiltration of CUI from the unclassified network of the DIB firm. DoD would bulk-purchase CSTs that are too costly for many small and some medium-sized DIB firms to afford. In exchange for these tools and services, DIB firms would take steps to secure CUI on their unclassified networks.

The DCP2 offers benefits to DIB firms, DoD, and the Intelligence Community, as it would enable real-time threat intelligence to be collected and synthesized across the DIB in ways currently not possible. The SOC would generate and disseminate alerts to DIB firms to secure and improve the monitoring of their networks from external and internal threats. The DCP2 would provide disadvantaged yet important DIB firms access to CSTs in a way that incentivizes their participation and protects DIB firms' CUI and the CUI in their supply chains. Similarly, the DCP2 would provide DoD with real-time insight into the cyber health of the DIB and help identify and respond to cyber threats. The DCP2 would not replace the proposed CMMC. The DCP2 is designed to complement the CMMC and better position DIB firms to comply with NIST SP 800-171 guidance.

We recognize that the DCP2 would impose significant costs on the government, costs that some could argue should instead be borne by private industry, given that private firms will benefit in many ways from the CSTs provided by DoD. However, we argue that it is the U.S. government's responsibility to protect the DIB—DIB firms are under cyber attack by competent nation-states using significant resources that in many cases greatly exceed those available to DIB firms. Just as in other domains, private companies should be and are protected from criminal actors by law enforcement agencies (e.g., by local police departments or the Federal Bureau of Investigation [FBI]).[5] DIB firms are also entitled to some form of cybersecurity protection by the U.S. government, although providing such protection requires a partnership across public and private entities to be successful.

---

[4] The commercial SOC would be run by a vetted and cleared U.S. cybersecurity service provider. This service would be paid for by DoD.

[5] However, it is important to note that DoD, not the FBI, has lead responsibility for protecting DIB firms under U.S. law. If the FBI were to take the lead role in protecting DIB firms from cyber attack, this would introduce significant legal concerns and complications.

Two Options for the Security Operations Center

We recommend that the SOC for the DCP2 be isolated from all other cybersecurity analysis centers in DoD, the Intelligence Community, and law enforcement agencies. If the FBI were to take the lead in protecting the unclassified networks of DIB firms, it could potentially expose firm employees and corporate officers to unrelated law enforcement investigations and actions, which potentially could violate DIB firms' Fourth Amendment rights. We recommend that significant legal protections be offered to participating DIB firms as part of the DCP2 to minimize any chance that additional liabilities would be incurred by DIB firms or their employees.

We offer two options for implementing the SOC for the DCP2. In Option A, DoD would play a direct role in real-time cyber defense of DIB firms. To facilitate this, the DIB SOC would be directly connected to the unclassified networks of DIB firms. The DIB SOC would provide sanitized dynamic intelligence, alerts, and recommended responses to DIB firms and, in turn, would deliver cybersecurity data collected by CSTs to the DIB SOC.

In Option B, DoD would play an indirect role in real-time cyber defense of DIB firms, and a commercial SOC would be directly connected to the unclassified networks of DIB firms. The commercial SOC would provide dynamic intelligence, alerts, and recommended responses to DIB firms and, in turn, they would deliver CST data to the commercial SOC. The commercial SOC would also be connected to the DIB SOC, which would aggregate data from multiple commercial SOCs to monitor the health of the DIB. Option B would reduce the probability that privately owned CUI or sensitive DIB firm data would be inadvertently sent to DoD. Option B may also present fewer legal concerns to some DIB firms. However, it may be more expensive, as it would require more SOCs to be established and operated.

DIB Cyber Protection Program Options: Moving the Unclassified Networks of Defense Industrial Base Firms to a Defense Industrial Base Cloud

The most cost-effective option for implementing the DCP2 may be based on cloud computing capabilities. In this option, DoD would establish a DIB cloud that could be used by DIB firms for computing and storage of unclassified data. DIB firms would move their unclassified networks into the DIB cloud. If such a DIB cloud were implemented, the CUI held by DIB firms would no longer be stored on their premises; instead, it would be stored and processed only in the DIB cloud.

The cloud service provider (CSP) would provide a secure enclave in a commercial cloud and a standardized set of computer system resources (CSRs) for that enclave for the DCP2. The DCP2 would provide a DIB cloud virtual machine (VM) and container repository with standardized VM and container images that can be used by DIB firms. The CSP would assume responsibility for patching and updating the cloud infrastructure used by DIB member firms. DoD would also establish and maintain a DIB cloud metadata service.

DIB firms that participate in the DCP2 would be provided a standardized set of secured CSRs in their own security enclaves. The security enclaves of individual firms would be separate from one another and would provide hard security boundaries between DIB firms to prevent the unauthorized flow of CUI and proprietary information.

The on-premises network would consist of thin-client or thick-client machines configured to prevent local storage of corporate data. No CUI would be stored in the on-premises network.

Maintaining Supply Chain Confidentiality and DIB Cyber Protection Program Eligibility

Our approach respects and maintains the confidentiality and proprietary nature of DIB contractor supply chains. To implement the DCP2, DoD will have to determine a company's eligibility for CSTs and cybersecurity services in a way that does not compromise supply chain relationships of DIB firms to the DCP2 government program managers. One of the long-standing challenges of administering a program like the proposed DCP2 is ensuring that DCP2 resources are made available to all DIB firms with CUI. Some smaller firms may not currently know they are part of a DoD supply chain. Such firms may provide a critical technology to an intermediate-level contractor that wishes to hide the source of the critical technology from DoD prime contractors for competitive reasons.

Proposed DFARS Flow-Down Clause for Controlled Unclassified Information

For the DCP2 to be successful, it would have to preserve supply chain confidentiality while fostering greater DIB transparency and verification of which firms have CUI. We propose that a new DFARS clause be included in DoD contracts that requires DIB firms to declare whether the DIB firm holds CUI and whether its immediate subcontractors hold CUI. The DIB firm would be required to declare the type of CUI it holds that is pertinent to DoD. The exact nature of the CUI would not have to be disclosed to the government, but the existence of the CUI would. DoD would use this information to make a decision on whether the DIB firm or any of its subcontractors are eligible for the DCP2.

This new DFARS clause would flow down to subcontractors, meaning that the contracts between the prime contractor and its subcontractor would contain this clause. The flow-down of the DFARS contract clause would require the subcontractors to disclose to the government whether they hold any CUI pertinent to DoD. This would ensure that DoD would obtain at least two CUI declarations for a subcontractor: one from the subcontractor itself and one from the DIB firm above it in the supply chain for the DoD program. In this way, DoD would be able to obtain a comprehensive list of DIB firms with CUI that should be eligible for the DCP2. DoD would use this information to grant DCP2 membership to DIB firms. This approach would preserve the confidentiality of the supply chain based on the DFARS flow-down clause, because CUI declarations would be made directly only to DoD.

Next Steps

Additional work will be required to determine the detailed cost of the proposed SOC for the DCP2. In addition, CST licensing costs and models should be explored and should include economies of scale and pricing options. It will not be a reasonable economic proposition to offer CSTs to every DIB firm. Thresholds and limits will have to be established to determine the number of CSTs paid for by DoD, and multiple CST subsidy models should be explored.

DIB firms may be ambivalent about sharing network and application metadata and anonymized account behavior data with DoD. However, the cybersecurity industry has developed CSTs that sanitize cyber artifacts, which can be used to detect anomalous behavior without sending the internal contents of files to an external SOC. Further research on CSTs is required to confirm these claims and to determine when additional data-sanitization tools will be needed to preserve the privacy and Fourth Amendment rights of DIB firms and employees.

Finally, it will be important to manage the cost of the DCP2. Only DIB firms that hold important CUI and provide DoD with critical defense-related technologies should be eligible to receive the full benefits of the program. Smaller firms that supply mostly commodity-related items to defense programs may not be eligible. A parametric cost analysis should be conducted that estimates the cost of the DCP2 and varies the number of DIB firms with important CUI.

Acknowledgments

We wish to thank several individuals at RAND for their support. First, we thank RAND-Initiated Research leadership, Susan Marquis and Howard Shatz, for their guidance, as well as several within the RAND National Security Research Division (NSRD) and the NSRD Cyber Intelligence Policy Center—Jack Riley, Rich Girven, and Sina Beaghley. We also thank Caolionn O'Connell and Thomas Donahue, the reviewers of this report, for their helpful comments and suggestions. Our report was much improved by their thorough reviews. Lastly, we thank Silas Dustin for his expert assistance throughout the course of our project.

Our research is informed by discussions we had with information security and cybersecurity professionals within RAND and other U.S. cybersecurity companies and tool providers identified through widely respected industry reports. These outside companies include CrowdStrike, Cylance, Fidelis Cybersecurity, FireEye, and Forcepoint. We also met with select DIB firm stakeholders and RAND subject-matter experts throughout our effort to inform the development of our cybersecurity framework. We are grateful for the time spent and insights shared during these meetings—our work is much improved because of these companies' inputs.

Of course, any errors in this report are the authors' alone.

Abbreviations

1FA	single-factor authentication
2FA	two-factor authentication
AI	artificial intelligence
APT	advanced persistent threat
CAC	Common Access Card
CFR	Code of Federal Regulations
CIO	chief information officer
CISA	Cybersecurity and Infrastructure Security Agency
CMF	Content Monitoring and Filtering
CMMC	Cybersecurity Maturity Model Certification
CSOC	cybersecurity operations center
CSP	cloud service provider
CSR	computer system resource
CST	cybersecurity tool
CTI	controlled technical information
CUI	controlled unclassified information
DAR	data at rest
DARPA	Defense Advanced Research Projects Agency
DCP2	DIB Cyber Protection Program
DFARS	Defense Federal Acquisition Regulation Supplement
DIB	defense industrial base
DIM	data in motion
DIU	data in use
DIUx	Defense Innovation Unit Experimental
DLP	data loss prevention
DoD	U.S. Department of Defense

DoDI	Department of Defense Instruction
DVD	digital versatile disc
EDR	endpoint detection and response tool
EO	Executive Order
FedRAMP	Federal Risk and Authorization Management Program
FOUO	For Official Use Only
FTE	full-time equivalent
FTP	File Transfer Protocol
HTTP	HyperText Transfer Protocol
HTTPS	HyperText Transfer Protocol Secure
HV	high-value
IDS	intrusion detection system
IP	intellectual property
IP	Internet Protocol
IT	information technology
MAC	Media Access Control
MFA	multifactor authentication
MITM	man-in-the-middle
MV	moderate-value
NAICS	North American Industry Classification System
NIST	National Institute of Standards and Technology
NTA	network traffic analysis
PII	personally identifiable information
POTUS	president of the United States
SIEM	security information and event management
SMC	security monitoring center
SOC	security operations center
SP	Special Publication
USB	universal serial bus
USC	U.S. Code
VM	virtual machine
VPN	virtual private network

CHAPTER ONE

Introduction

The defense industrial base (DIB) is under attack. China and other foreign actors are stealing large amounts of sensitive data, trade secrets, and intellectual property (IP) every day from DIB firms—contributing to the erosion of the DIB and potentially harming U.S. military capabilities and future U.S. military operations.[1] Several recent reports have detailed China's activities in cyberspace,[2] and the 2019 U.S. Department of Defense (DoD) report *Military and Security Developments Involving the People's Republic of China 2019* noted that "China uses its cyber capabilities to not only support intelligence collection against U.S. diplomatic, economic, academic, and DIB sectors, but also to exfiltrate sensitive information from the DIB to gain military advantage."[3] Furthermore, attacks to government unclassified networks have seen a dramatic increase.[4] In 2018, DoD Defense Innovation Unit Experimental (DIUx) estimated the cost of U.S. IP exfiltrated from U.S. firms by nation-state cyber actors to be $300 billion per year.[5] The U.S. Secretary of the Navy in 2018 noted, "attacks on our networks are not new, but attempts to steal critical information are increasing in both severity and sophistication."[6]

---

1 In this report, we define *sensitive data* as trade secrets, IP, confidential contract terms and pricing, proprietary supply chain data (including firm names and products), and personally identifiable information (PII) of DIB firm corporate officers, employees and/or associates.

2 See, for example, Michael Brown and Pavneet Singh, *China's Technology Transfer Strategy: How Chinese Investments in Emerging Technology Enable A Strategic Competitor to Access the Crown Jewels of U.S. Innovation*, U.S. Department of Defense, Defense Innovation Unit Experimental (DIUx), January 2018.

3 U.S. Department of Defense, *Annual Report to Congress: Military and Security Developments Involving the People's Republic of China 2019*, Washington, D.C., May 2019, p. 65.

4 U.S. Department of Defense, "Cybersecurity Challenges: Protecting DoD's Unclassified Information," PowerPoint briefing, Washington, D.C., August 15, 2018, slide 3.

5 Brown and Singh, 2018.

6 Gordon Lubold and Dustin Volz, "Chinese Hackers Breach U.S. Navy Contractors," *Wall Street Journal*, December 14, 2018.

DoD is taking steps to better secure systems against cyber threats, but struggles remain.[7] Most well-established protections in place focus on classified networks, whereas the unclassified networks have become an attractive "backdoor" entrance for adversaries seeking access to cutting-edge research and development efforts. Previous RAND research found that DoD lacks a comprehensive strategy for protecting the unclassified networks of DIB firms. Investigations after significant cyber attacks found that "it's a matter of trust and hope to secure the systems of their contractors and subcontractors"[8] and "subcontractors across the entire military were lagging behind in cybersecurity and frequently suffered breaches that affected other branches."[9]

DoD has increased regulations and introduced new security controls,[10] but this approach may be insufficient—DIB firms cannot keep up. Cybersecurity is necessary but also expensive; cybersecurity tools (CSTs) can be costly and require skilled, highly trained professionals to use, and the required CSTs may not be affordable for many smaller DIB firms. Furthermore, the regulatory environment is complex and challenging to navigate, even for large firms with robust cybersecurity teams and tools.

Because of these challenges, more U.S. government assistance is needed to defend the DIB from cyber intrusions. Advanced persistent threats (APTs) will continue to penetrate the unclassified networks of DIB firms. Even the best commercial software products have vulnerabilities against the sophisticated capabilities of nation-state actors, and perimeter defenses are insufficient to protect against ATPs. In 2019, the director of the National Security Agency called for the public and private sectors to unite against cybersecurity threats:

> Expecting the private sector to literally withstand the focused efforts of entire nation states that are working in a very synchronized strategy way to attempt to gain advantage, I don't think that's realistic.[11]

Objective and Approach

This report offers a way ahead that would involve a more active role for DoD that goes beyond the current regulatory push. Our objective is to better secure unclassi-

[7] Lubold and Volz, 2018.

[8] Tom Bossert, former Homeland Security Advisor to U.S. President Donald Trump, as quoted in Lubold and Volz, 2018.

[9] Bossert, as quoted in Lubold and Volz, 2018.

[10] DoD activities to improve cybersecurity range from regulations to voluntary programs. See U.S. Department of Defense, "Cybersecurity Challenges: Protecting DoD's Unclassified Information," PowerPoint briefing, Washington, D.C., August 15, 2018, slide 4.

[11] Quoted in Bradley Barth, "Former NSA Director: Public and Private Sectors Must Unite Against Cyberattacks," *SC Magazine*, March 7, 2019.

fied defense networks through the establishment and implementation of a cybersecurity program to strengthen the protections of these networks. Our approach includes several steps. First, we define the DIB, characterize the population of DIB firms, and identify the population most vulnerable to cyber intrusion and most disadvantaged in terms of being able to afford adequate protections. Next, we discuss the current DoD approach for protecting unclassified networks from cyber attacks, including the legal protections in place for DIB firms. We also assess the cost of cybersecurity: How much do firms currently spend on cybersecurity, from the tools to the workforce, and how much do we estimate they should be spending? We then review current DIB cyber protection tools offered by leading cybersecurity firms. Finally, we develop a more robust DIB cyber protection program that includes

- **improved monitoring and real-time health** of the DIB
- **improved cybersecurity** for firms that cannot afford the CSTs and professional staff needed
- **data protections** to prevent the disclosure of sensitive corporate information from DIB firms to DoD, prevent sensitive supply chain information to be disclosed across the DIB, and prevent the exfiltration of DIB sensitive data to adversaries
- **legal protections** for DIB firms, to minimize the liability that DIB firms might have if the information they provide to the government is used in unanticipated ways.

Organization of This Report

In the next chapter, we define the DIB and what we mean by small, medium, and large firms, and we identify the key elements that we focus on in our research. We also provide an estimated breakdown of the DIB based on firm size and revenue. Chapter Three discusses current DoD protections for the DIB, shortfalls of the current approach, the legal landscape, and implications for DIB firms. Chapter Four discusses the current cost and state of cybersecurity using cybersecurity budget estimates, information technology (IT) budget estimates, and cybersecurity professional salary estimates. We develop an estimate for the recommended cybersecurity budget for small- and medium-sized DIB firms and compare that with current estimates. We end Chapter Four with a discussion of implications for the DIB. In Chapter Five, we describe cybersecurity tools typically used by DIB firms, including network access control, network defenses, vulnerability scanning and software patching, security information and event management (SIEM), email security, data filtering, and data loss prevention (DLP). In Chapter Six, we introduce our alternative DIB cybersecurity protection framework, with a discussion of legal protections and implications for DIB firms. In Chapter Seven, we summarize the findings and recommendations of our report.

This report also has several appendixes: Appendix A provides detailed network diagrams for our cyber protection framework; Appendix B describes cybersecurity tools from select cybersecurity firms; and in Appendix C we discuss DLP tools.

CHAPTER TWO

Defining the Defense Industrial Base

In this chapter, we define key DIB terms, describe the composition of the DIB, and identify the portion of the DIB we focus on in this study.

What Is the Defense Industrial Base?

The DIB is the set of private and public firms (from small to large companies) that provide defense industrial capabilities. Defense industrial capabilities are "the skills and knowledge, processes, facilities, and equipment needed to design, develop, manufacture, repair, and support DoD products and their necessary subsystems and components."[1] A 2018 DoD report to President Donald Trump in response to Executive Order (EO) 13806[2] on strengthening the manufacturing and the defense industrial supply chain further characterized the DIB as having two parts: (1) the domestic manufacturing and defense industrial base and (2) the global manufacturing and defense industrial base.[3]

As shown in Figure 2.1, the domestic base comprises producers of goods and services from small, medium, large firms. This base is further split into the private sector, which has some of the prime system integrators, and the organic industrial base, which includes government-owned, government-operated entities and government-owned, contractor-operated entities. The global manufacturing base consists of these enterprises in other countries, some of which the United States has formal relationships with and others it does not. Together, the domestic and global base provide defense industrial capabilities across a range of sectors shown along the bottom of the figure—from

1 Department of Defense Instruction 5000.60, *Defense Industrial Base Assessments*, Washington, D.C.: U.S. Department of Defense, July 18, 2014.

2 Executive Order 13806, *Assessing and Strengthening the Manufacturing and Defense Industrial Base and Supply Chain Resiliency of the United States*, Washington, D.C.: The White House, July 21, 2017.

3 U.S. Department of Defense, *Assessing and Strengthening the Manufacturing and Defense Industrial Base and Supply Chain Resiliency of the United States: Report to President Donald J. Trump by the Interagency Task Force in Fulfillment of Executive Order 13806*, Washington, D.C., September 2018.

Figure 2.1
DoD's Characterization of the Defense Industrial Base

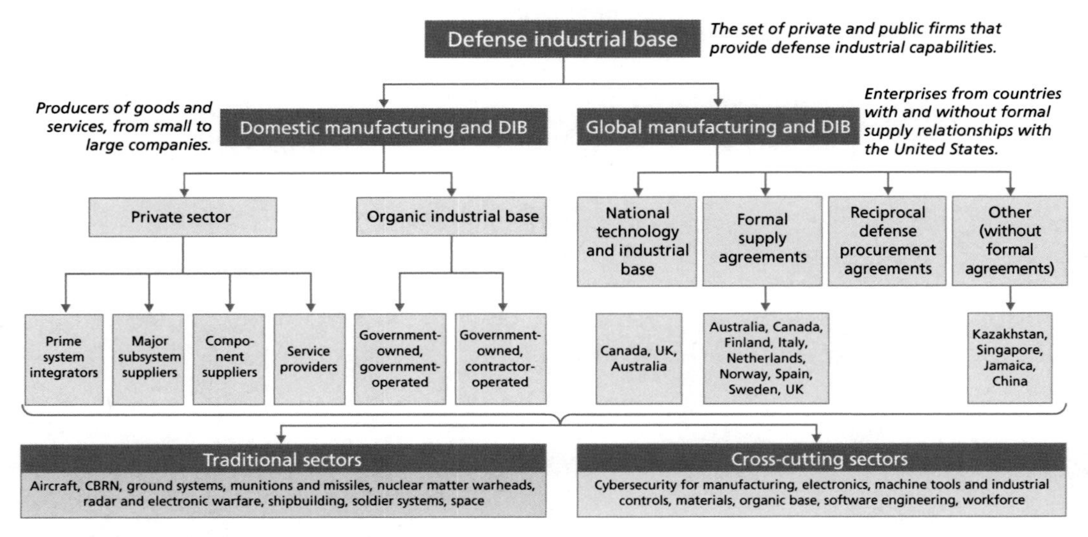

SOURCE: Derived from U.S. Department of Defense, 2018.
NOTE: CBRN = chemical, biological, radiological, and nuclear.

traditional (e.g., aircraft, ground systems) to more cross-cutting sectors (e.g., materials, software engineering).

For our research, we build on DoD's characterization of the DIB and focus on key elements of the DIB ecosystem within with domestic manufacturing and defense industrial base. In particular, we focus on small- and medium-sized firms in the private sector and including research and development organizations.

The red-outlined boxes in the figure indicate the focus of our study. Within the domestic base, our private-sector focus is on small- and medium-sized tech-sector firms conducting research for DoD. We also expand DoD's definition to include research and development within the domestic base—organizations conducting research and development for DoD, from academia, to federally funded research and development centers and other companies (Defense Advanced Research Projects Agency [DARPA] contractors).

The United States has a set of guidelines for what it considers small in terms of firm size, but there are no set definitions for medium and large firms.[4] Even with the guidelines, there is variation across the U.S. government for how *small* is defined. Academia and international standards provide other guidelines, as well.[5] For the purposes

---

[4] U.S. Small Business Association size standards can be found at U.S. Small Business Administration, "Table of Size Standards," webpage, August 19, 2019.

[5] For example, Ohio State University's National Center for the Middle Market defines medium-sized companies as those with a revenue between $10 million and $1 billion (See National Center for the Middle Market,

Figure 2.2
Study Focus: Key Elements of the DIB Ecosystem

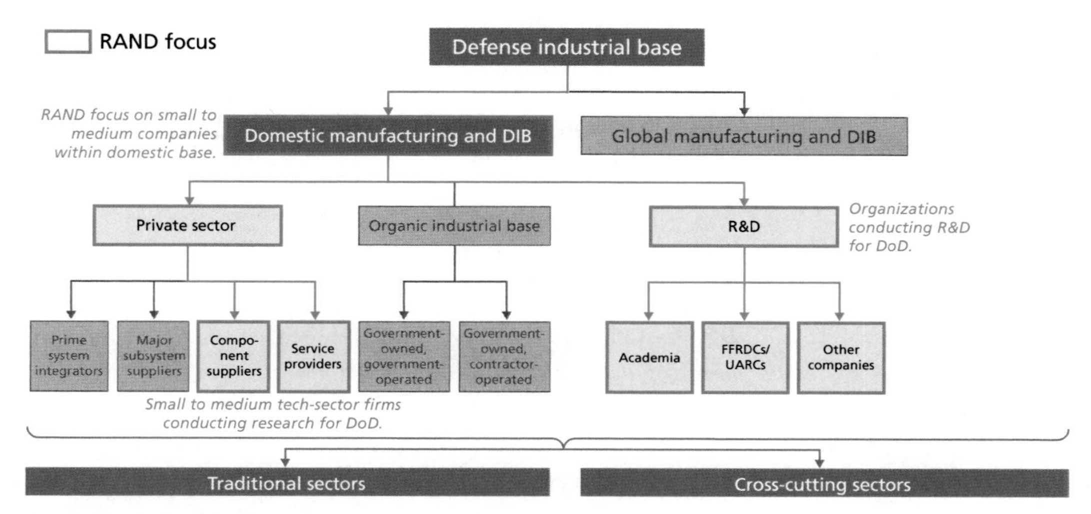

SOURCE: Derived from U.S. Department of Defense, 2018.
NOTE: UARC = university-affiliated research center.

of our study, small DIB firms are those with less than $100 million in annual revenue, medium firms are those with between $100 million and $500 million in annual revenue, and large firms are those with $500 million in annual revenue. Here are some examples as of fiscal year 2018:

- **Small DIB firms:** MaXentric Technologies, a DARPA contractor employing over 50 people, with an annual revenue of $5 million–$10 million;[6] and First RF Corp, a DARPA contractor employing approximately 100 people, with an annual revenue of $24 million.
- **Medium DIB firms:** Georgia Tech Research Corporation, part of the Georgia Institute of Technology, with a DoD revenue of approximately $393 million; and the Microsoft Corporation, a commercial high-technology firm, with a DoD revenue of approximately $400 million.
- **Large DIB firms:** Boeing, a prime contractor within the aircraft sector, employing 137,000 people, with a total annual revenue of $101 billion and DoD rev-

undated) In comparison, the Organisation for Economic Co-operation and Development defines small as a revenue less than 10 million euros, medium as between 10 million and 50 million euros, and large as over 50 million euros (Organisation for Economic Co-operation and Development, "Glossary of Statistical Terms," webpage, last updated on December 2, 2005; Organisation for Economic Co-operation and Development, "Data: Enterprises by Business Size [Indicator]," webpage, undated).

[6] D&B Hoovers, "MaXentric Technologies, LLC," webpage, undated; Manta, "MaXentric Technologies, LLC," webpage, undated; Glassdoor, "MaXentric Technologies," webpage, undated; D&B Hoovers, "First Rf Corporation," webpage, undated; Glassdoor, "First RF," webpage, undated.

enue of approximately $27 billion;[7] Lockheed Martin, a prime contractor within the aircraft, electronic, radar, and electronic warfare sector, employing 105,000 people, with a total annual revenue of $53.8 billion and DoD revenue of approximately $39 billion; and Northrop Grumman, a prime contractor within the aircraft, C4 [command, control, communications, and computers], electronics, radar, and electronic warfare sector, employing 85,000 people with a total annual revenue of $25.8 billion and DoD revenue of approximately $11 billion.[8]

How Big Is the Defense Industrial Base?

As discussed earlier, the DIB is made up of small, medium, and large firms. Here, we will further define *small*, *medium*, and *large* and provide an estimate of how many DIB firms there are in each category.

Top defense companies with large revenues are dominant players in the DIB but make up only a small percentage of overall DIB firms. However, there is no official DIB data set showing a list or breakdown of the DIB by size, partly because of supply chain sensitivities. Details about the composition of DoD supply chains, in particular the small firms in the lowest tier, are not available. Because the composition of individual firm supply chains is considered to be proprietary by many prime contractors and smaller suppliers, a complete list of DIB firms is not available to DoD decisionmakers. Our research and the approach we put forward later in this report respect the confidentiality of supply chains and supply chain memberships. To develop a surrogate snapshot of the DIB, we show firms in the United States by revenue using the most recent census data and North American Industry Classification System (NAICS) codes (collected in 2012 and released in 2015).[9] As shown in the top circle in Figure 2.3, over 99 percent of U.S. firms—as represented by the blue shading in the circle—have a revenue of less than $100 million,[10] and so are what we characterize as small firms. The circle in the bottom portion of the figure breaks down the remaining 0.39 percent, which are medium- to large-sized firms with revenue greater than $100 million.

This finding is consistent with data from the Census Bureau's "Annual Survey of Entrepreneurs," which noted that in 2016 there were 5.6 million employer firms: Firms with fewer than 500 workers accounted for 99.7 percent, firms with fewer than 100

---

[7] Federal Procurement Data System—Next Generation, "Top 100 Contractors Report," data sheet, 2018.

[8] Federal Procurement Data System—Next Generation, 2018; U.S. Department of Defense, *Fiscal Year 2017: Annual Industrial Capabilities*, Washington, D.C., March 2018.

[9] U.S. Census Bureau, "Number of Firms, Number of Establishments, Employment, Annual Payroll, and Estimated Receipts by Large Enterprise Receipt Sizes for the United States, NAICS Sectors: 2012," 2012 County Business Patterns and 2012 Economic Census, Washington D.C., 2012.

[10] We define small as less than $100 million in annual revenue, medium as $100 million to $500 million, and large as more than $500 million.

Figure 2.3
Breakdown of U.S. Firms by Revenue (as of 2012)

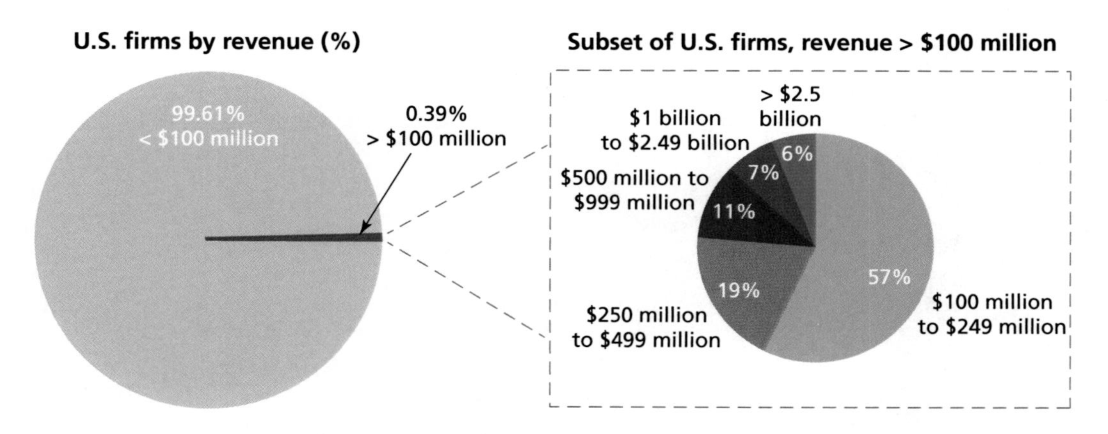

SOURCE: RAND analysis based on U.S. Census Bureau, 2012.

workers accounted for 98.2 percent, and firms with fewer than 20 workers accounted for 89.0 percent.[11]

Figure 2.4 shows the U.S. economy firm size distribution by revenue and full-time equivalent (FTE) employees.

We use the NAICS data to determine the firm size distribution by revenue and FTE for three layers: revenue greater than $500 million (large firms), revenue between $100 million and $500 million (medium firms), and revenue less than $100 million (small firms). Large firms, indicated by the top (green) layer, represent only about 5,000 firms yet have the greatest number of employees per firm. Medium firms, indicated by the middle layer (red) layer, have on average 911 FTEs. Finally, small firms, shown by the bottom (blue) layer, represent over 99 percent of overall firms and have an average of 11 FTEs.

We argue that the size distribution of DIB supply chains does not differ significantly from the size distribution of businesses in the overall U.S. economy. Therefore, we use the NAICS code U.S. economy firm size percentages as a proxy to better understand the distribution of firms across the DIB. We use the Federal Procurement Data System to estimate the number of DoD contractors with a DoD revenue (annual dollars obligated by DoD to the firm) greater than $500 million, which is 63 firms as of fiscal year 2018.[12] This is not based on total firm revenue and does not include revenue from non-DoD contracts. We then apply the percentages reflecting the distribution

---

[11] Small Business and Entrepreneurship Council, "Facts & Data on Small Business and Entrepreneurship," webpage, undated.

[12] Federal Procurement Data System—Next Generation, 2018.

Figure 2.4
U.S. Economy Firm Size Distribution

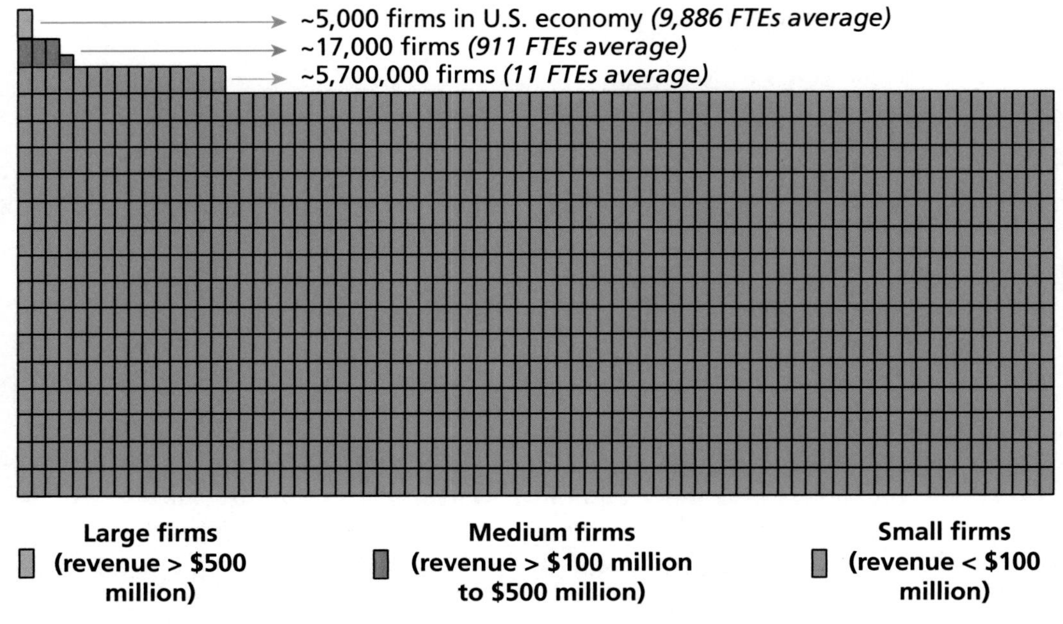

SOURCE: Federal Procurement Data System—Next Generation, 2018.

of the U.S. economy to understand the DIB distribution for the bottom two layers. Figure 2.5 shows the estimated DIB firm size distribution based on these calculations.

As shown in the figure, we estimate the total number of DIB firms to be approximately 72,000. We note that others have estimated this number to be higher, and closer to approximately 100,000; however, we have not been able to find any breakdown to explain these higher numbers.

Figure 2.5
Estimated DIB Firm Size Distribution

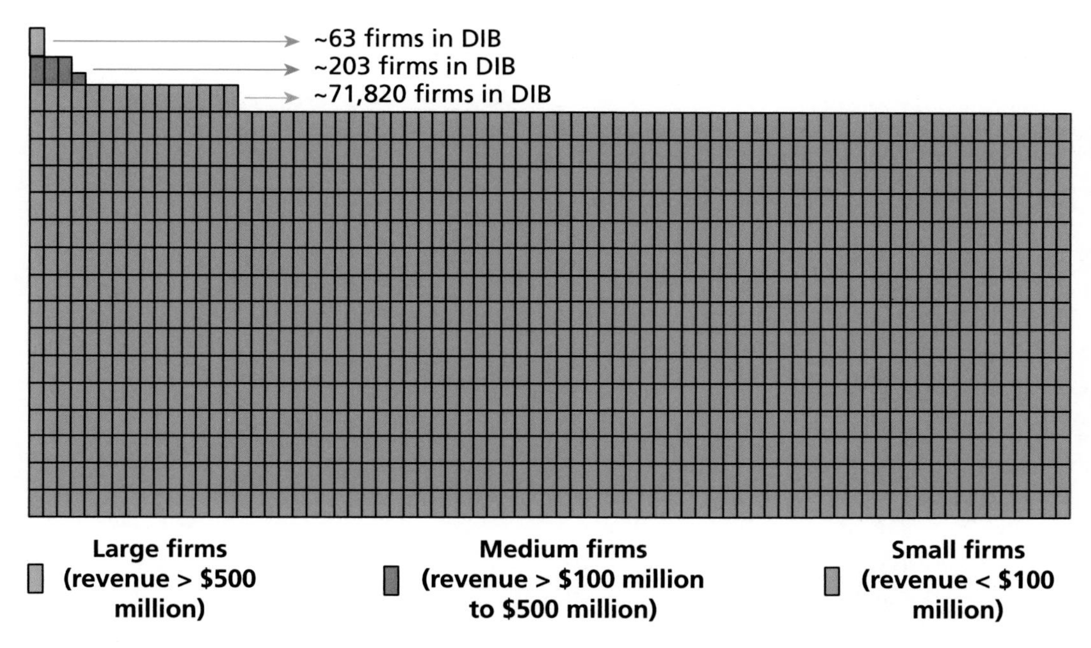

SOURCE: Federal Procurement Data System—Next Generation, 2018.

CHAPTER THREE

Current Defense Industrial Base Protections

As described in Chapter One, DIB firms face relentless cyber attacks by foreign strategic competitors attempting to steal trade secrets, advanced technologies, and IP, including design information of key military systems. In the past several years, DoD has realized the extent of this threat and devised a set of policies to address these issues. In this chapter, we review DoD's current approach for protecting these types of sensitive information. We also discuss shortfalls of the current approach, describe the legal landscape for DIB firms, and close with implications for these firms.

The U.S. Government's Definition of Trade Secrets and Intellectual Property

In 2013, then–Under Secretary for Acquisition Technology and Logistics Frank Kendall stated during a Senate hearing that while he felt U.S. classified technical data were well protected, he was less confident in the United States' ability to protect sensitive unclassified data.[1] Kendall spoke of sensitive design data, which typically resides on a defense contractor's unclassified network and is proprietary.[2] Since 2013, China has attacked U.S. academic universities conducting research for DoD, the Navy, and the DIB more broadly. Reports noted that the Pentagon writ large "faces mounting challenges" in protecting its systems from cyber threats."[3] In these attacks, China has been targeting trade, economic, and military secrets specifically associated with critical technology. The data have been characterized as "sensitive" but not classified, illustrating the vulnerability and value of data residing on unclassified networks both within

[1] Sydney J. Friedberg Jr., "Top Official Admits F-35 Stealth Fighter Secrets Stolen," *Breaking Defense,* June 20, 2013.

[2] "Theft of F-35 Design Data Is Helping U.S. Adversaries—Pentagon," Reuters, June 19, 2013.

[3] Justin Katz, "Alarmed by Lack of Ongoing Research, Navy Cyber Group Seeks Defensive Tech from Industry," *Inside Defense,* July 15, 2019.

DoD and across the DIB.[4] These sensitive design data make up a subset of what the United States terms controlled unclassified information (CUI).

Until recently, DoD lacked a full appreciation for how much valuable information resides on unclassified networks, particularly technical design data for advanced technologies and capabilities. This realization partly served as the impetus for the issuance of EO 13556, *Controlled Unclassified Information*, in 2010. EO 13556 identified information that did not reach the threshold of classification but should still be protected[5]—hence the term *CUI*.[6] Our research also focuses on controlled technical information (CTI), which is a subcategory of CUI that includes trade secrets, information about advanced technologies, and IP.[7]

Either the government or a DIB firm can own CTI. If government-owned, it typically means that the government at some point purchased the technical rights to a program or system from a DIB firm. However, often the DIB firm owns CTI. When a DIB firms owns CTI, the valuable information that DoD wishes to protect resides only on a DIB firm's unclassified networks, protected only by whatever protections the firm has in place.

DoD's Current Approach to Protecting Controlled Unclassified Information

In response to EO 13556, DoD established two processes for protecting CUI (Figure 3.1). First, DoD implemented procedures for protecting CUI on DoD internal networks. Second, DoD designed a set of regulations and processes for defense contractors to protect this subset of information on their unclassified networks. Our focus is on the latter set of practices, within defense contractor unclassified networks.

DoD designed a three-part approach for defense contractors to responsibly handle CUI on their unclassified networks. The first two aspects of this approach represent the regulatory and compliance elements. The third component is voluntary in nature and somewhat restrictive. We explore each part of the DoD approach in greater detail and discuss its advantages and disadvantages. Table 3.1 outlines the various documents we discuss in this section, in addition to those relevant to DoD's approach to

---

4 Dustin Volz, "Chinese Hackers Target Universities in Pursuit of Maritime Military Secrets," *Wall Street Journal*, March 5, 2019; Gordon Lubold and Dustin Volz, "Navy, Industry Partners Are 'Under Cyber Siege' by Chinese Hackers, Review Asserts," *Wall Street Journal*, March 12, 2019.

5 Executive Order 13556, *Controlled Unclassified Information*, Washington, D.C.: The White House, November 4, 2010.

6 Executive Order 13556, 2010.

7 National Archives, "CUI Category: Controlled Technical Information," webpage, undated.

Figure 3.1
DoD Approach to Protecting Controlled Unclassified Information

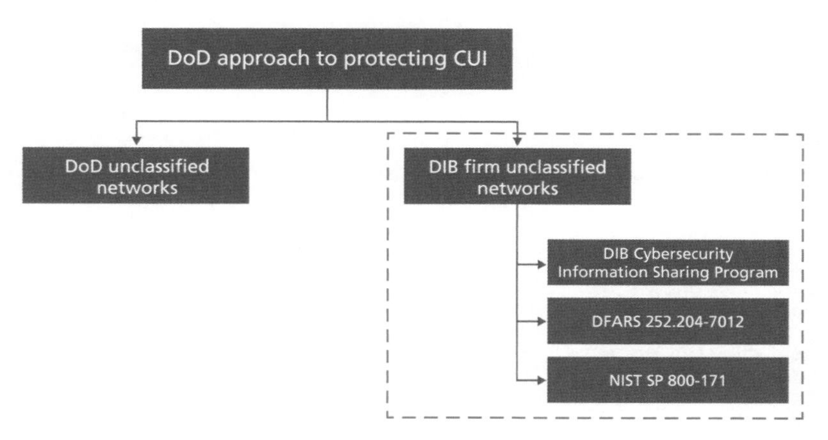

NOTES: DFARS = Defense Federal Acquisition Regulation Supplement; NIST SP 800-171 = National Institute of Standards and Technology Special Publication 800-171 (Ross et al., 2019a).

the cybersecurity of DIB and DoD unclassified networks. First, we explore the voluntary component—an information-sharing website open only to those with a Common Access Card (CAC).

DIB Cybersecurity Information Sharing Program

The DIB Cybersecurity Information Sharing Program was designed as a voluntary program intended to provide an outlet for defense contractors to report incidents and find useful threat intelligence. Outlined in 32 CFR Part 236, the DIB Cybersecurity Information Sharing Program attempted to encourage greater participation by placing it outside of the mandatory aspects of DoD's DIB cybersecurity activities. The program provides a digital portal for defense contractors to access cyber threat information from other DIB firms. It is meant to be a collaborative environment for DoD and DIB parties. However, in practice, the voluntary program has proven more exclusive than inclusive. The program was modeled after Cybersecurity and Infrastructure Security Agency's threat information–sharing website, but to be part of that program, a firm has to have public key infrastructure (PKI) certificates or a CAC. Many DIB contractors have neither.

In practice, the DIB Cybersecurity Information Sharing Program helps those with access by providing useful unclassified and For Official Use Only (FOUO) information on cyber threats but is exclusive in nature for many smaller companies inside the DIB. The voluntary program cannot reach the whole DIB. Additionally, the portal provides only part of the available threat intelligence picture. It likely does not provide a complete set of dynamic threat intelligence, as it may exclude the latest real-time threat intelligence and does not include classified information. Therefore, although a

Table 3.1
Policy and Legal Documents

Date	Document Title	Type	Description	Lead Agency
May 27, 2009	Classified Information and Controlled Unclassified Information	Presidential Memorandum	Calls for a review of CUI procedures and a CUI interagency task force.	POTUS
Sep 2010	Department of Defense Information Network (DODIN) Transport	Department of Defense Instruction (DoDI 8010.01)	Outlines the policy, procedures, responsibility to the DoD information network.	DoD
Nov 4, 2010	Controlled Unclassified Information	Executive Order (EO 13556)	Establishes CUI category and approach for handling.	POTUS
Sep 14, 2016	Controlled Unclassified Information	Federal Regulation (32 CFR Part 2002)	Legal framework for CUI.	
Oct 2016	Disclosure of Information Act	DFARS (252.204-7000)	Outlines the guidelines for sharing CUI.	DoD
Sep 2017	Implementation of DFARS Clause 252.204-7012, Safeguarding Covered Defense Information and Cyber Incident Reporting	Memorandum for DoD Leadership	Memorandum to assist DoD acquisition personnel in ensuring that contractors implement the NIST 800-171 standards.	DoD
Sep 2018	Implementation of Enhanced Security Controls on Select Defense Industrial Base Partner Networks	Memorandum for Distribution	Requests for all future Navy contracts to have a security plan and data requirement list.	Navy
Nov 2018	Guidance for Assessing Compliance and Enhancing Protections Required by DFARS Clause 252.204-7012, Safeguarding Covered Defense Information and Cyber Incident Reporting	Memorandum for DoD Leadership	More information and guidance on how to incorporate cybersecurity protections into the procurement process. Directed at DoD acquisition personnel.	DoD
Revised as of June 2019	Safeguarding Covered Defense Information and Cyber Incident Reporting	DFARS (252.204-7012)	Outlines requirements for defense contractors processing covered defense information, or CUI.	DoD
June 2019	NIST SP 800-171 (Ross et al., 2019a)	Cybersecurity Guidelines	Provides the requirements for protecting CUI.	NIST
In revision	NIST SP 800-171B (Ross et al., 2019b)	Cybersecurity Guidelines	Supplement to NIST SP 800-171.	NIST
April 2013	NIST SP 800-53 (Joint Task Force Transformation Initiative, 2013)	Information Security and Privacy Guidelines	Provides guidelines on the security and privacy of information. Supplement to the Federal Information Security Management Act.	NIST

useful contribution to DoD's current approach, the DIB Cybersecurity Information Sharing Program may not be a complete solution to the issues facing unclassified networks today from adversaries.[8]

DFARS 252.204-7012

The second component of DoD's current approach involves the Defense Federal Acquisition Regulation Supplement (DFARS) 252.204-7012, "Safeguarding Covered Defense Information and Cyber Incident Reporting," a document that broadly outlines the desired contractual requirements of DIB firms for the cybersecurity of their unclassified networks processing CUI.[9]

The DFARS also provides useful definitions to various key terms, including *adequate security*, *controlled technical information*, and *covered defense information*. We list these definitions below to clarify how the regulatory document defines certain categories of information and the avenues to protect them:

- **Adequate security:** protective measures that are commensurate with the consequences and probability of loss, misuse, or unauthorized access to, or modification of information.
- **Contractor attributional/proprietary information:** information that identifies the contractor(s), whether directly or indirectly, by the grouping of information that can be traced back to the contractor(s) (e.g., program description, facility locations), personally identifiable information, as well as trade secrets, commercial or financial information, or other commercially sensitive information that is not customarily shared outside of the company.
- **Controlled technical information:** technical information with military or space application that is subject to controls on the access, use, reproduction, modification, performance, display, release, disclosure, or dissemination.
- **Covered contractor information system:** an unclassified information system that is owned, operated by or for, a contractor and that processes, stores, or transmits covered defense information.
- **Covered defense information:** unclassified controlled technical information or other information, as described in the CUI Registry, that requires safeguarding or dissemination controls pursuant to and consistent with law, regulations, and governmentwide policies and is
  - marked or otherwise identified in the contract, task order, or delivery order and provided to the contractor by or on behalf of DoD in support of the performance of the contract; or

---

[8] See the DIB Cybersecurity Information Sharing Program website for more information (DIBNet, undated).

[9] DFARS 252.204-7012, "Safeguarding Covered Defense Information and Cyber Incident Reporting," October 2016.

  - collected, developed, received, transmitted, used, or stored by or on behalf of the contractor in support of the performance of the contract.
- **Cyber incident:** actions taken through the use of computer networks that result in a compromise or an actual or potentially adverse effect on an information system and/or the information residing therein.

A compliance-based document, DFARS 252.204-7012 outlines three primary requirements for contractors: (1) report cyber incidents, (2) provide adequate security, and (3) flow-down clause. We discuss these three elements below.

Cyber Incident Reporting

Section d of DFARS 252.204-7012 describes what defense contractors must do after they realize a cyber incident has occurred. These actions include creating a report, submitting it to a DoD portal, and adhering to any follow-on actions by DoD, such as a damage assessment. For cyber incident reporting, the DFARS provides a list of criteria, with the following most pertinent to our study:

- Report cyber incident to DoD through the DIBNet portal.
- Create an incident report.
- Submit malicious software to the DoD Cyber Crime Center (DC3).
- Include the DFARS clause in all subcontracts that will involve covered defense information (CUI).

The cyber incident reporting section of the DFARS notes that a cyber incident does not mean a contractor lacks "adequate security." Many small firms learn they have been subject of a cyber attack only after the FBI contacts them regarding the intrusion. However, the FBI may not know a small firm is a defense contractor. The new DFARS language closes this gap. The next section explores what DoD has deemed adequate security.

Adequate Security

In general, DFARS 252.204-7012 defines *adequate security* to mean adhering to the security controls of the National Institute of Standards and Technology (NIST) Special Publication (SP) 800-171. DFARS distinguishes between information systems controlled or operated by the government—and thus subject to government security requirements—and information systems and networks owned and operated by the defense contractor. For defense contractor networks, the regulation requires the firm to implement security controls laid out in NIST SP 800-171. Additionally, if the contractor wishes to deviate from the special publication, it must send a formal request to the contracting officer. DFARS 252.204-7012 states that these requests will eventually reach the DoD chief information officer (CIO), who will then conduct an adjudication process.

The current DoD approach permits DoD contractors to self-certify that they comply with NIST SP 800-171. Some members of Congress have criticized the lack of a rigorous certification process in current DFARS language. For this reason, DoD is now developing a new approach that will require an external third party to certify that a DIB firm is complying with NIST guidance. Although fulfilling the adequate security criterion within the DFARS document may seem straightforward, the security controls within NIST SP 800-171 are complex, costly, and particularly difficult for smaller DIB firms to navigate and implement.

Flow-Down Clause

Section m of DFARS 252.204-7012 outlines the responsibilities and requirements for subcontracts of defense contractors, emphasizing that the entire DFARS 252.204-7012 clause be included in subcontracts if they meet certain criteria. Subcontracts that require the inclusion of DFARS 252.204-7012 include those "for operationally critical support, or for which subcontract performance will involve covered defense information." Therefore, all subcontractors that meet this criterion are required to implement NIST SP 800-171. The prime contractor is responsible for denying the subcontractor access to CUI if the contractor refuses to follow DFARS 252.204-7012.

Additionally, this section of the DFARS assumes that CUI flows from the prime contractors, yet not all relevant trade secrets and IP may be owned by the prime contractor. Because of this top-down approach, the regulation will inevitably not cover all CTI, as many subcontractors remain in business because of their trade secrets. Many subcontractors have knowledge of how to make products and IP on software code they own that prime contractors wish to leverage. This information also needs to be protected but is overlooked in the current flow-down clause.

NIST SP 800-171

NIST SP 800-171, *Protecting Controlled Unclassified Information in Nonfederal Systems and Organizations*, is a guidance document that has been adopted as the cornerstone of a compliance-based approach to DIB cybersecurity. DFARS 252.204-7012 requires defense contractors to implement NIST SP 800-171's more than 100 security controls. The first revision of NIST SP 800-171 was published in December 2016 and updated in June 2018,[10] with the second revision still in draft form.[11] NIST SP 800-171B, a sep-

[10] Ron Ross, Kelley Dempsey, Patrick Viscuso, Mark Riddle, and Gary Guissanie, *Protecting Controlled Unclassified Information in Nonfederal Systems and Organizations*, NIST Special Publication 800-171, Revision 1, Washington, D.C.: U.S. Department of Commerce, National Institute of Standards and Technology, December 2016, updated June 2018.

[11] Ron Ross, Victoria Pillitteri, Kelley Dempsey, Mark Riddle, and Gary Guissanie, *Protecting Controlled Unclassified Information in Nonfederal Systems and Organizations,* Draft NIST Special Publication 800-171, Revision 2, Washington, D.C.: U.S. Department of Commerce, National Institute of Standards and Technology, June 2019a.

arate document, serves as a supplement to 800-171, giving more details on enhanced security, specifically against APTs.[12] NIST SP 800-171 outlines 14 categories of controls for protecting CUI but does not contain an approved product list or banned product lists for critical network functions or endpoint applications. To adequately satisfy these categories—such as access control, incident response, and maintenance—requires sophisticated and expensive cybersecurity tools and applications. Some of these functions that require such tools and applications include

- security monitoring
- logs capture, log data correlation and analysis
- security event alerting
- IT device inventory monitoring and control
- time synchronization of across all devices on the network
- application blacklisting
- control and monitor all user software applications.

To carry out these tasks, firms will need trained cybersecurity professionals and a sufficient budget (discussed later in this report) for cybersecurity applications. The skilled cybersecurity professionals provide the expertise to install, maintain, and use CSTs to perform the bulleted functions above, required by NIST 800-171. Both of these—skilled staff and proper tools—require thousands to millions of dollars, depending on the size of a firm. Further, DoD does not currently provide any explicit assistance or resources to firms to comply with NIST 800-171.

Two key issues with NIST SP 800-171 are its lack of resources to assist DIB firms in implementing the controls and lack of a certification process to ensure firms adhere to the controls. A third, and perhaps most telling, problem is that, as of July 2019, no defense contractor has complied with 100 percent of the security controls. DIB firms were to implement NIST SP 800-171 by December 31, 2017, but a May 2019 Sera-Brynn study found that all, and especially smaller companies, had failed to comply.[13]

Public Criticism of NIST SP 800-171 and 800-171B

The May 2019 study conducted by Sera-Brynn showed that DIB firms find difficulty in implementing the standards outlined in NIST SP 800-171, giving rise to public

---

[12] The latest draft version of SP 800-171B was published in June 2019 (Ron Ross, Victoria Pillitteri, Gary Guissanie, Ryan Wagner, Richard Graubart, and Deborah Bodeau, *Protecting Controlled Unclassified Information in Nonfederal Systems and Organizations: Enhanced Security Requirements for Critical Programs and High Value Assets*, Draft NIST Special Publication 800-171B, June 2019b.

[13] Justin Doubleday, "New Report Finds Defense Contractors Struggling with Cybersecurity Requirements," *Inside Defense,* May 21, 2019a.

attention and criticism of the standards document.[14] Not surprisingly, the report found that the smallest DIB companies had the greatest difficulty with compliance, larger firms had the highest percentage of compliance (though far from 100 percent), and medium-sized firms had similar levels of adherence to the controls. These results demonstrate that compliance with NIST SP 800-171 will continue to be a challenge for most DIB firms. Ultimately, the findings from this report generated a larger discourse on the challenge DoD continues to face in securing the unclassified networks within its supply chain, and that the current approach may need a significant revision.

In August 2019, the wireless industry expressed public concern regarding the costs of implementing NIST SP 800-171 security controls. The trade association for the wireless communications industry, CTIA, openly called for NIST to "reconsider its assessment of the costs of compliance with NIST SP 800-171B, which will likely be substantial."[15] NIST, as of August 2019, delayed releasing the latest version of NIST SP 800-171B, pending Office of Management and Budget review of NIST SP 800-53, a document that explores that statutory responsibilities of Federal Information Security Management Act (FISMA).[16] The wireless industry has been forthright in stating that if NIST does not delay the release of 800-171B to include industry feedback and account for the high cost of implementation, it will undermine its mission of protecting government information.[17]

Additionally, critics have pointed out that DoD does have any certification process or mechanism to ensure that DIB firms comply with NIST 800-171. A July 2019 DoD Inspector General report found not only that DIB firms continue to struggle with NIST 800-171 compliance, but also that the government does not have any verification or certification process in place to facilitate compliance.[18] Though NIST SP 800-171 is intended to improve the security of DIB firm networks, it has been overlooked by many firms because of the lack of a certification requirement.[19]

In response to criticism over the lack of DoD certification process for NIST SP 800-171, DoD plans to revise its approach to incorporate a maturity model that uses

---

14 Sera-Brynn, *Reality Check: Defense Industry's Implementation of NIST SP 800-171: Keen Insights from Certified Cybersecurity Assessors,* May 2019.

15 Rick Weber, "Wireless Industry Warns of Costs, Other Concerns from NIST Cyber Standards for Defense Contractors," *Inside Defense,* August 23, 2019.

16 NIST Special Publication 800-53 consists of a series of guidance and security controls for the privacy and security of Federal information and systems. 800-53 was initiated as an effort expand upon the statutory requirements outlined in Public Law 107-347, Federal Information Security Management Act (FISMA), December 12, 2002.

17 Weber, 2013.

18 Inspector General, U.S. Department of Defense, *Audit of Protection of DoD Controlled Unclassified Information on Contractor-Owned Networks and Systems,* Washington, D.C., July 23, 2019.

19 Doubleday, "Pentagon to Require New Cybersecurity 'Certification' from Defense Contractors," *Inside Defense,* June 6, 2019b.

a third party to certify compliance. DoD labeled the new approach as DoD Cybersecurity Maturity Model Certification (CMMC).[20] The model includes 18 domains based on cybersecurity best practices. The capabilities within the model refer to processes and practices mapped to five levels of maturity. Level 1 maturity corresponds approximately to what we categorize as the current level of cybersecurity practice in small DIB firms. Level 3 maturity, according to the CMMC, includes some application and consideration of the NIST SP 800-171 security controls. Level 5 includes advanced cybersecurity practices and tools. These include cryptographically secure multifactor authentication (MFA), DLP, and SIEM capabilities.[21] These will be discussed in Chapter Six. While DoD recognizes that the current approach does not properly secure its supply chain and CTI, the CMMC approach may not be the correct fix either. In August 2019, the Under Secretary of Defense for Acquisition and Sustainment, in discussing the CMMC, stated that DoD was "extremely concerned" about supporting small businesses, which often do not have the resources to implement strict cybersecurity controls on their networks.[22]

Shortfalls of DoD's Current Approach to Protecting Controlled Unclassified Information

DoD's current approach to protecting CTI on DIB unclassified networks does not account for the complexity of the security controls, expertise required to implement and maintain compliance with them, and the implications of not having an organization or capability to certify compliance with the NIST SP 800-171 controls. In short, the requirements in NIST SP 800-171 and related documents are not feasible for small DIB firms because of the extent of the requirements and the high cost of the tools and staff required to properly protect an unclassified network that has DoD CTI. The current approach fails to adequately account for the complexity and cost imposed on firms working to comply and does not provide a usable framework for DIB firms to follow to improve the cybersecurity of their unclassified networks.

---

[20] Justin Doubleday, "Pentagon to Require New Cybersecurity 'Certification' from Defense Contractors," *Inside Defense*, June 6, 2019b.

[21] U.S. Department of Defense, "Cybersecurity Maturity Model Certification (CMMC): Draft CMMC Model Rev 0.4 Release and Request for Feedback," briefing, September 2019.

[22] Quoted in Tony Bertuca and Justin Doubleday, "Pentagon Reveals New Acquisition Initiatives to Block China," *Inside Defense*, August 26, 2019.

Legal Landscape

In this section, we provide an overview of the legal landscape for DIB companies that share cyber threat information with the federal government. In December 2015, Congress passed and the President signed the Cybersecurity Act of 2015.[23] Title I of the Cybersecurity Act, titled the Cybersecurity Information Sharing Act (CISA),[24] provides authority for cybersecurity sharing between the private sector and the federal government.[25]

CISA provides four important authorities for detecting and sharing cybersecurity threats:

1. "a specific, but broad grant of authority to private sector entities to conduct monitoring, for cybersecurity purposes, of their own systems and the systems of customers that provide authorization and written consent for such monitoring"[26]
2. a broad grant of authority to private sector entities to share cyber threat indicators and defensive measures with one another and with the federal government for cybersecurity purposes[27]
3. a broad grant of authority to private sector entities to operate defensive measures on their own information systems, as well as on the systems of their customers that provide appropriate written authorization and consent for such operations[28]
4. specific liability protections for private entities that conduct such monitoring activities.[29]

There are two aspects of CISA that are particularly important to our report. First, the minimization requirement of the legislation, which requires private-sector entities (such as DIB firms), prior to sharing any information concerning cyber threat indicators, to first review those indicators to assess whether they contain any personally identifiable information (PII) that is not directly related to a cybersecurity threat. If

---

23 Enacted as part of Public Law 114-113, Consolidated Appropriations Act, December 18, 2015; 129 Stat. 2242.

24 The CISA enacted as part of Public Law 114-113 has no relation to the relatively new U.S. Department of Homeland Security Cybersecurity and Infrastructure Security Agency, which uses the same acronym.

25 CISA also provides increased authority for cybersecurity information sharing between and among state, local, tribal, and territorial governments and the federal government, in addition to the private sector. See Section 104(c)(1) of CISA.

26 CISA, Section 104 (a)(1).

27 See CISA, Section 104(c)1).

28 See CISA, Section 104(b)(1).

29 CISA's Section 106(a) states: "No cause of action shall lie or be maintained in any court against any private entity, and such action shall be promptly dismissed, for the monitoring of an information system and information under Section 104 (a) that is conducted in accordance with this title."

such information is detected, it must be removed prior to sharing the information.[30] Second, CISA permits a private-sector entity to "implement and utilize a technical capability configured to remove any information not directly related to a cybersecurity threat that the non-Federal entity knows at the time of sharing to be personal information of a specific individual or information that identifies a specific individual."[31] One CISA expert noted that "imposing a minimization-like requirement, somewhat narrow though it may be, will likely make companies less likely to share in the first instance, at least until the market develops CISA-compliant sharing systems or mechanisms that employ a technical capability along the lines authorized by statute."[32]

In June 2016, the U.S. Department of Homeland Security (DHS) and the U.S. Department of Justice issued *Guidance to Assist Non-Federal Entities to Share Cyber Threat Indicators and Defensive Measures with Federal Entities Under the Cybersecurity Information Sharing Act of 2015*. The guidance defines *cyber threat indicator* and provides numerous examples of what could be considered one. The guidance states, "Effectively, the only information that can be shared under the Act is information that is directly related to and necessary to identify or describe a cybersecurity threat."[33] The guidance describes the process that nonfederal entities (such as DIB firms) may use for sharing cyber threat indicators and defensive measures with federal entities, as provided in Section 104 (c) of CISA. The guidance describes the DHS program and process through which cyber threat indicators and defensive measures may be shared. However, the guidance also clarifies that, "Consistent with CISA, non-federal entities may also share cyber threat indicators and defensive measures with federal entities through means other than the Federal government's capability and process operated by DHS."[34]

Concerns about the sharing of cyber threat information pursuant to CISA have been raised by privacy advocates and technology companies, among other stakeholders. These concerns can be grouped into four categories:

- the broad definition of data to be monitored, collected, and shared with other private entities or federal government[35]

---

[30] See CISA, Section 104(d)(2).

[31] See CISA, Section 104(d)(2)(A)–(B).

[32] Jamil N. Jaffer, "Carrots and Sticks in Cyberspace: Addressing Key Issues in the Cybersecurity Information Sharing Act of 2015," *South Carolina Law Review*, Vol. 67, 2016.

[33] U.S. Department of Homeland Security and U.S. Department of Justice, "Guidance to Assist Non-Federal Entities to Share Cyber Threat Indicators and Defensive Measures with Federal Entities under the Cybersecurity Information Sharing Act of 2015," June 15, 2016, p. 5.

[34] U.S. Department of Homeland Security and U.S. Department of Justice, 2016, p. 15.

[35] John Heidenreich, "The Privacy Issues Presented by the Cybersecurity Sharing Act," *North Dakota Law Review*, Vol. 91, 2015, pp. 395–410.

- lack of clear guidelines for the collection, sharing, and retention of data by private entities and the federal government[36]
- government use of data obtained by private entities via CISA[37]
- risk of the institution of a federal government surveillance program.[38]

For example, critics have suggested that CISA "would give the government sweeping new powers to spy on Americans in the name of protecting them from hackers."[39] The *Washington Post* reported that Senator Ron Wyden of Oregon stated, "Sharing information about cybersecurity threats is a worthy goal. . . . Yet if you share more information without strong privacy protections, millions of Americans will say, 'That is not a cybersecurity bill. It is a surveillance bill.'"[40] In addition, CISA has been criticized for sharing information without privacy protections, failing to provide protection for individual privacy rights, and violating civil liberties.[41] Critics have pointed out that CISA does not provide a good mechanism to deal with wrongfully disclosed personal information, and that the statute does not provide a private right of action for violations of CISA provisions.[42] Another concern that has been raised about CISA is that information received by the federal government can be used in legal proceedings by any federal agency or department for purposes unrelated to cybersecurity.[43] Similarly, privacy advocates have warned that liability protections granted to companies may result in the oversharing of customer data with the government, contributing to the government's ability to conduct surveillance activities.[44]

---

36 See, e.g., Heidenreich, 2015.

37 "The larger problem is that CISA authorizes the government to collect a huge amount of data without a warrant or probable cause" (Heidenreich, 2015).

38 Brian Fung, "Apple and Dropbox Say They Don't Support a Key Cybersecurity Bill, Days Before a Crucial Vote," *Washington Post*, October 20, 2015; Damian Paletta and Daisuke Wakabayashi,, "Apple Piles on as Senate Debates Cyber Bill; Apple Joins Twitter in Opposing Information Sharing Legislation," *Wall Street Journal*, October 21, 2015

39 Fung, 2015.

40 Quoted in Fung, 2015.

41 "CISA Security Bill Passes Senate with Privacy Flaws Unfixed," *Wired*, October 27, 2015; Abigail Tracy, "The Problems Experts and Privacy Advocates Have with the Senate's Cybersecurity Bill," *Forbes*, October 29, 2015.

42 See, e.g., Jay P. Kesan and Carol M. Hayes, "Bugs in the Market: Creating A Legitimate, Transparent, and Vendor-Focused Market for Software Vulnerabilities," *Arizona Law Review*, Vol. 58, No. 3, 2016; Mark Jaycox, "EFF Opposes Cybersecurity Bill Added to Congressional End of Year Budget Package," Electronic Frontier Foundation, December 18, 2015.

43 Specifically, CISA's Section 105(d)(5)(A) permits any agency of the Federal government to use information obtained pursuant to CISA to prevent or mitigate specific threats of death, serious bodily injury, serious economic crime, fraud and identity theft, espionage and censorship, and the protection of trade secrets.

44 See, e.g., Tracy, 2015; CISA's Section 105(d)(3)(A)–(B) provide that information collected by the Federal Government pursuant to CISA "shall be" "exempt from disclosure" and "withheld, without discretion, from the public."

Another statute that pertains to our report is the Cybersecurity Enhancement Act of 2014. In Title I, "Public-Private Collaboration on Cybersecurity," the legislation amended the National Institute of Standards and Technology Act (15 USC 272[c]) to require that NIST "facilitate and support the development of a voluntary, consensus-based, industry led set of standards, guidelines, best practices, methodologies, procedures and processes to cost-effectively reduce cyber risks to critical infrastructure."[45] Additionally, the legislation requires NIST to "identify a prioritized, flexible, repeatable, performance-based, and cost-effective approach, including information security measures and controls, that may be voluntarily adopted by owners and operators of critical infrastructure to help them identify, assess and manage cyber risks."[46] As explained previously in the report, as of July 2019, no defense contractor (including prime contractors) has been able to fully comply with NIST SP 800-171. Another piece of cybersecurity legislation, the Federal Information Security Modernization Act of 2014, which amended the Federal Information Security Management Act of 2002 (FISMA), pertains primarily to cybersecurity practices of the federal government, not private-sector entities such as DIB firms.

Implications for Defense Industrial Base Firms

CISA will have the greatest impact on DIB firms that wish to share cyber threat information with the federal government. The most obvious challenge for private firms will be to effectively "minimize" information concerning PII prior to providing it to the federal government. However, CISA provides a number of safe harbors for DIB firms that share information according to the legislation's requirements:

- **No liability for cyber threat monitoring activities** on a company's own information system, or the information systems of customers, if appropriate consent and authorization is obtained.[47]
- **No civil liability for information sharing** with the federal government concerning cyber threats or defensive measures if the sharing is conducted in accordance with CISA.[48]
- **No waiver of any applicable privileges or protections** as a result of information sharing pursuant to CISA. The threat or defensive measures information shared will not be subject to state or federal Freedom of Information Act laws; it can be designated as the proprietary information of the private company that shares it.[49]

---

[45] Public Law 113-274, Cybersecurity Enhancement Act of 2014, December 18, 2014, Section 101(a)(2)(15).

[46] Public Law 113-274, Cybersecurity Enhancement Act of 2014, December 18, 2014, Section 101(b)(e)(1)(A)(iii).

[47] CISA, Sections. 104 (a)(1) and 106(a).

[48] CISA, Section 106(b).

[49] CISA, Section 105(d)(1).

- **No antitrust liability for two or more companies that share threat information** or provide cybersecurity assistance. "It shall not be considered a violation of any provision of antitrust laws for private companies to exchange or provide a cyber threat indicator or defensive measure."[50]
- **No duty is imposed on private companies to share cybersecurity information**, and there is no requirement that a company act on the receipt of cyber threat information. A federal entity may not "condition the award of any grant, contract, or purchase" on the reciprocal exchange of cyber threat information. [51]
- **No new government regulations may be created from cyber threat or defensive measures information** provided pursuant to CISA except regulations concerning how information systems can prevent or mitigate cybersecurity threats.[52]

Although CISA legislation explicitly states that "cyber threat indicators and defensive measures provided to the Federal Government under this title shall not be used to . . . regulate the lawful activities of any non-Federal entity or any activities taken by a non-Federal entity pursuant to mandatory standards," it also provides a major exception: "Cyber threat indicators and defensive measures provided to the Federal government under this title may, consistent with Federal or State regulatory authority specifically relating to the prevention or mitigation of cybersecurity threats to information systems, *inform the development or implementation of regulations* relating to such information systems."[53] It may be a disincentive for DIB firms to provide threat information to the federal government if they think it could result in new, more stringent, and probably expensive cybersecurity regulation. The cost of cybersecurity protection for DIB firms will be discussed in detail in the next chapter.

[50] CISA, Section 104(e).

[51] CISA, Sections 108(i) and (h).

[52] CISA, Section 105(d)(5)(D)(i).

[53] CISA, Section 105(d)(5)(D)(i).

CHAPTER FOUR

Current Cost and State of Cybersecurity

In the previous chapter, we discussed DoD's current approach to protecting DIB firms. However, the question is, do DIB firms have the resources to comply with these requirements, which would include implementing complex security controls, purchasing CSTs, and maintaining a competent cybersecurity workforce? In this chapter, we identify the amount that firms typically spend on cybersecurity for their internal networks. We then compare this with the amount of cybersecurity spending that experts at IT firms suggest is needed to protect internal information resources. For this analysis, we reviewed the cybersecurity spending literature and best practices for general commercial firms and estimate IT and cybersecurity spending of typical DIB firms based on comparable spending for different types of commercial firms. Our cost analysis shows that most small and many medium-sized DIB firms are incapable of fully complying with the recommended cybersecurity measures because they cannot afford to, leaving their networks and the DIB writ large vulnerable to cyber attack.

Cybersecurity Budget Estimates

A firm's cybersecurity budget is often a small component of the larger IT budget. A 2018 survey from *CIO Magazine* found that just 25 percent of companies spend 10–20 percent of their IT budgets on security, with over half of those surveyed spending less than 10 percent.[1] Determining an average cybersecurity budget is difficult given that spending varies across firms in different industries and that different "average" spending numbers can vary nearly 300 percent among sources.[2] In 2016–207, PwC, Gartner, and Forrester estimated cybersecurity spending as being 3.7 percent, 5.9 percent, and 10 percent of IT spending, respectively, for a typical commercial firm, which indicates the wide range of cybersecurity spending estimates from these three

[1] Josh Fruhlinger, "The State of IT Security, 2018," *CIO*, May 29, 2018.

[2] Alex Asen, Walter Bohmayr, Stefan Deutscher, Marcial Gonzalez, and David Mkrtchian, "Are You Spending Enough on Cybersecurity?" Boston Consulting Group, February 20, 2019.

respected consulting groups.[3] According to Bromium, an average large business (one with more than 2,000 employees) spends $16.7 million annually on security software and services.[4] For this analysis, we adopt Gartner's midrange 6 percent estimate as the average spending on cybersecurity as a proportion of IT spending. With this number in mind, it is important to note that many organizations do not feel sufficiently secure against cybersecurity threats. A 2019 survey of 850 global organizations with 10–1,000 employees found that 52 percent feel helpless to defend themselves from new forms of cyber attacks.[5]

Estimated Information Technology Budgets of Defense Industrial Base Firms

The ability of a firm to create and maintain an efficient cybersecurity system is dependent on its revenue, the size and complexity of the network that must be protected, and the value of the information held in the network. In its 2012 *IT Enterprise* report, Gartner published average IT spending as a percentage of revenue for a range of industries. These IT budget numbers vary widely, from 1.1 percent to 7.6 percent of total firm revenue.[6] Financial firms spend more on cybersecurity than firms in most other industries, for example. It should be noted that the Gartner data does not include a specific entry for DIB firms. The Gartner estimates of IT spending as a percentage of the total revenue is also dated, based on data from 2012 or earlier. Nevertheless, one can select a small number of industries from the Gartner data for which IT spending estimates are available, and average these to develop an estimate for a typical DIB firm. Using this approach, we estimate that the average IT budget of a DIB firm is 4.2 percent of total annual revenue.

A later report, drawing on Deloitte's 2016–2017 Global CIO Survey, concluded that the average percentage of total annual firm revenue spent on IT for all industries is 3.28 percent, again with banking and securities industries and business and professional services industries estimated to spend the greatest proportion of their revenue on IT, 7.61 percent and 5.82 percent, respectively.[7] A still more recent study published

[3] PwC, *The Global State of Information Security Survey*, London, UK, March 10, 2017; Gartner, *IT Key Metrics Data 2017*, Stamford, Conn., December 12, 2016; Forrester Research, *2017 Tech Budget Benchmark*, Cambridge, Mass.: Forrester Research, March 28, 2017.

[4] Bromium, Inc., *The Hidden Costs of Detect-to-Protect Security*, Cupertino, Calif., 2018.

[5] Vanson Bourne, *Underserved and Unprepared: The State of SMB Cyber Security in 2019*, Boston, Mass.: Continuum Managed Services, 2019.

[6] Jamie Guevara, Eric Stegman, and Linda Hall, *Gartner IT Key Metrics Data 2012: IT Enterprise Summary Report*, Stamford, Conn.: Gartner, 2012.

[7] Khalid Kark, Caroline Brown, and Anjali Shaikh, "Technology Budgets: From Value Preservation to Value Creation," *Deloitte Insights*, November 28, 2017.

by Computer Economics in 2019 found the 25th and 75th percentiles of IT spending for firms in the high-tech industry to be 2.6 percent and 4.7 percent.[8] We chose to use this most recent estimate for high-tech firms as a surrogate for a typical DIB firm with CUI. Using this upper estimate for high-tech industry firms in 2019, we estimate that DIB firms spend approximately 4.7 percent of their revenue on IT. This is comparatively higher than the average of all industry benchmarks, but, given the complexity and the high technology content of products made by DIB firms, we believe they are likely to spend more on IT than the average commercial firm. Also, this estimate of 4.7 percent agrees well with estimate arrived at using the older Gartner data, of 4.2 percent.

Cybersecurity Professional Salary Estimates

Providing a firm with effective cyber defenses requires more than just new CSTs. Cybersecurity threats continue to evolve rapidly. No single tool can protect the firm. Firms must have cybersecurity professionals on staff to manage the suite of CSTs installed on the network, to monitor the alerts and warnings generated by CSTs, analyze the artifacts generated by network components and endpoints, and to respond to cyber alerts and breaches. Even though new CSTs are more capable than older ones that required much more manual processing, they still generate data that require skilled professionals to review and interpret. Skilled cybersecurity professionals are in short supply. In 2018, the median pay for information security analysts was $98,350.[9] (Salaries for cybersecurity professionals can vary greatly, however, with a range between $45,000 to $150,000.) In addition to the take-home pay, firms compensate their employees through a range of benefits, including health insurance and contributions to retirement accounts. According to the Bureau of Labor Statistics, benefits constitute 33.6 percent of overall compensation for an employee in the information services sector in the private industry.[10] Based on the estimates of the average salary and cost of benefits, a single cybersecurity professional would cost an employer approximately $131,396 annually.

There appears to be a lack of consensus on cybersecurity personnel staffing recommendations or requirements, perhaps because the threat has been increasing and older staffing level estimates have been determined to be inadequate. Though a few industry estimates are available, these estimates vary significantly. A snapshot of cur-

---

[8] *Computer Economics*, "IT Spending as a Percentage of Revenue by Industry, Company Size, and Region," 2019.

[9] U.S. Department of Labor, Bureau of Labor Statistics, "Information Security Analysts," September 4, 2019.

[10] U.S. Department of Labor, Bureau of Labor Statistics, "Employer Costs for Employee Compensation—March 2019," June 18, 2019.

rent cybersecurity staffing levels at a small sample of DIB firms (shown in Table 4.1) confirms the wide variance of firms' cybersecurity FTE staff numbers as a function of firm revenue. Furthermore, some of the staffing numbers shown in Table 4.1 show that cybersecurity staffing levels are increasing substantially at some DIB firms.

The SANS Institute conducted a recent survey on the challenges small and medium-sized businesses face in ensuring the cybersecurity of their firms.[11] The median employee count of the 22 small and medium-sized businesses was 80 employees, and the mean number of cybersecurity staff was 1 employee, with one-third outsourcing their cybersecurity needs.[12] When asked how many cybersecurity staff the firm would like to have, the mean answer was two FTEs, or four FTEs for IT and cybersecurity.[13] McAfee conducted a study assessing the cybersecurity workforce in Australia and found that small and medium-sized businesses had an average of 9.7 cybersecurity employees and that companies with more than 500 employees had an average cybersecurity workforce of 11.4 people.[14] NuHarbor Security claims "a general rule is your security staff should be between 5–10% of your IT staff."[15]

Cybersecurity staffing numbers are often described as a percentage of the firm's entire staff or their IT staff, rather than the firm's revenues, which is how we categorize DIB firms in this report. Other studies also categorize the revenue of small and medium-sized businesses in different ways, using different firm revenue thresholds. We have attempted to align these cybersecurity staffing data from other studies using firm revenue numbers. Using this approach, we find that only the smallest of firms can securely function with just one cybersecurity professional. Larger firms will likely require more skilled cybersecurity professionals because of the increasing number of endpoints and network components that have to be monitored. We estimate that firms with revenue between $10 million and $20 million should have at least one cybersecurity FTE staff member and that firms with a revenue of $30 million to $50 million need to have two cybersecurity FTEs. At $100 million of revenue, firms should have about six cybersecurity employees; at around $300 million of revenue, firms should have ten cybersecurity FTEs. The number of cybersecurity staff needed by larger firms, with revenues of $300 million or more, increases more slowly after this threshold, consistent with the findings of the NuHarbor study cited earlier. We combine these cybersecurity staffing recommendations and convert them into predicted cybersecurity staffing costs

[11] Small and medium-sized businesses are defined as firms with fewer than 500 employees.

[12] Aric Asti, *Cyber Defense Challenges from the Small and Medium-Sized Business Perspective*, Bethesda, Md.: SANS Institute, 2019.

[13] Asti, 2019.

[14] McAfee, *Cybersecurity Talent Study: A Deep Dive into Australia's Cybersecurity Skills Gap*, Santa Clara, Calif., September 2018. McAfee's cybersecurity study defines small businesses as having fewer than 250 employees and medium-sized businesses having between 250 and 500 employees.

[15] NuHarbor Security, "Information Security Staffing Guide," March 05, 2019.

for DIB firms of differing sizes (in revenue). These data are illustrated by the red lines in Figures 4.1 and 4.2, shown later in this chapter.

Characteristics of a Small Sample of Defense Industrial Base Firms

Next, we compare these cybersecurity staffing and spending estimates with actual data from a small number of anonymous DIB firms. Through a trusted third party, we discussed cybersecurity issues with four anonymous DIB firms of various sizes (referred to as Firms A–D in Table 4.1). The sizes of these firms ranged from small (less than $100 million in revenue) to medium (roughly $350 million of revenue). Table 4.1 shows the results of and insights gleaned from these discussions.

Firm A and Firm D currently spend more than the previously mentioned estimates of firms' cybersecurity spending as a percentage of their IT budget. These spending numbers are in fact more aligned with recommended spending targets, as explained below. According to the firms we spoke with, several smaller DIB firms are already using DLP, two-factor authentication (2FA), and perhaps other advanced CSTs, while larger ones are just beginning to use such tools. Some smaller DIB firms have found it necessary to deploy DLP and 2FA to protect their IT and trade secrets. DLP capabilities can also help protect data from destructive malware, such as ransomware. However, we note that this is not a scientific survey and is based on discussions with four firms. The four firms also addressed their satisfaction with their cybersecurity staff and

Table 4.1
Characteristics of a Sample of DIB Firms

DIB Firm Characteristics	Firm A	Firm B	Firm C	Firm D
Annual revenue	~$240 million	$101 million–$250 million	$81 million–$100 million	~$350 million
Cybersecurity, FTE staff	12	5	9	14
DLP tools used to fingerprint sensitive data and prevent sensitive data loss	Yes	Yes	NA	No
2FA used to log on to company computers	Yes	Yes	Yes	Not 100%
2FA used to log on to company network	Yes	Yes	Yes	Yes
Percentage of IT budget spent on cybersecurity	35% (growing to 50%)	NA but growing	NA	37%

NOTE: NA used to indicate where information is not available. Firms B and C declined to provide an estimate of their cybersecurity spending, even as a ratio of their IT budgets. 2FA = two-factor authentication.

budget (Table 4.2). Only Firm A, with a revenue of around $240 million, felt that its cybersecurity budget and staff were sufficient.

Comparing the Estimated and Recommended Cybersecurity Budgets of Small and Medium-Sized DIB Firms

As previously mentioned, we estimate that a typical commercial firm currently spends, on average, 6 percent of its IT budget on cybersecurity. Although there is no widely accepted benchmark for cybersecurity spending, as needs, resources, and risk tolerance levels vary among firms, available recommended spending targets are over two and three times greater than the current 6 percent cybersecurity spending estimate. IBM recommends that companies with significant cybersecurity concerns spend 14 percent of their IT budgets on cybersecurity measures.[16] Data from Forrester suggest that if a company has been hacked it will spend 30 percent or more of its IT budget on cybersecurity.[17]

We estimate that, on average, a DIB firm should spend 22 percent of its IT budget on cybersecurity, which is the average of the 14 percent and 30 percent recommendations mentioned above. We feel that this target is acceptable, given that we assume that as many as half of the DIB may have been subject to cyber attack. Even before the introduction of the new DFARS 252 contract language, these DIB firms that have been victims of cyber attacks most likely will, on average, spend more on cybersecurity. Based on press reports of targeted military contractors, particularly for the Navy and Air Force; insights gleaned from interviews with DIB firms; and the independent Forrester estimate, we believe that the recommended spending level of 22 percent will be needed to address the threat. For the cost analysis, however, we will also estimate the impact of spending 14 percent of an IT budget on cybersecurity, which is a lower recommendation from IBM for firms with significant cybersecurity concerns.

Table 4.2
A Sample of DIB Firms' Satisfaction with Their Cybersecurity Resources

DIB Firms' Cybersecurity Resources	Firm A	Firm B	Firm C	Firm D
Do you have enough cybersecurity staff to meet your needs?	Yes	No	No	NA
Is your firm's cybersecurity budget adequate?	Yes	No	Yes	NA

[16] Katharina Gerberding, "Cybersecurity Budgeting 101: How to Optimize Your Security Spend for Maximum ROI," Hitachi Security Systems, June 26, 2018.

[17] Jeff Pollard, *Security Budgets 2019: The Year of Services Arrives*, Cambridge, Mass.: Forrester, December 17, 2018.

Cyber Security Spending by Small Defense Industrial Base Firms

Figure 4.1 and 4.2 display, for small and medium-sized DIB firms, respectively, the differences between estimated current and recommended overall cybersecurity spending, recommended spending on cybersecurity professionals, and estimated IT budgets. We estimate that there are approximately 71,820 DIB firms in the United States with a revenue of less than $100 million (see Figure 2.5). Figure 4.1 shows the budget challenges small DIB firms likely have in funding their cybersecurity needs.

The two recommended cybersecurity spending levels are indicated in Figure 4.1 as yellow (14 percent) and purple (22 percent) lines, both of which are greater than the current estimated spending on cybersecurity, indicated by lighter gray bar for each firm size. The recommended level of spending on cybersecurity staff is indicated by the red line and is derived using the sources and method described earlier. To have the minimum recommended of number of cybersecurity professionals on staff, small DIB firms would have to devote all or almost all of their recommended cybersecurity budget to staffing. The figure shows that recommended spending on cybersecurity staff (red line) overlaps the 22 percent overall cybersecurity spending (purple line) for firms with revenue of $30 million or less, and recommended spending on staff is greater than the 14 percent overall spending (yellow line) for all small DIB firms. The financial requirement for cybersecurity professionals alone is far greater than the recommended 5.9 percent of IT budgets on cybersecurity small firms, and it is greater than current estimated cybersecurity spending for all firms with revenue less than $100 million.

Figure 4.1 also shows our estimate for how much Firm C spends on cybersecurity staff based on the Bureau of Labor Statistics data referred to earlier. This esti-

Figure 4.1
Estimated and Recommended Cybersecurity Budgets of Small DIB Firms

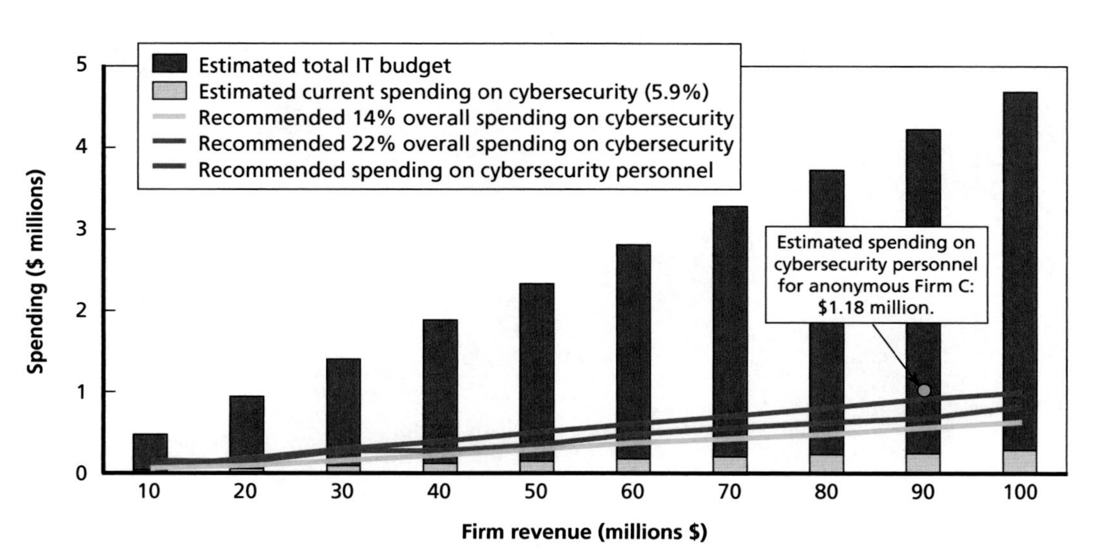

mate shows that Firm C's spending on cybersecurity staff is probably more than even the higher recommended 22 percent of IT budget spending (the purple line). Using the cybersecurity professional labor cost estimates described above, we estimate that Firm C spends $1.18 million on its nine cybersecurity employees, or nine FTEs. This number is approximately 25 percent of the estimated IT budget of Firm C.

Using two independent methods and different sources, we show that small DIB firms likely do not have the financial resources needed to procure and maintain a robust cyber defense against sophisticated nation-state adversaries. Small DIB firms, including even the most profitable within this category, are unlikely to have enough money for all the CSTs they need if they hire and retain the number of recommended cybersecurity personnel for firms of their size. Another independent study shows that the typical small firm cannot pay for cybersecurity professionals and that the mean number of cybersecurity professionals employed by a typical small firm is only one FTE.[18] If a firm of this size invests its resources in cybersecurity personnel, it is unlikely to have enough resources for the tools to secure its unclassified networks. DIB firms in this group are likely struggling either to have sufficient cybersecurity professionals or to maintain a full suite of CSTs tools, and they likely cannot have both at the same time.

Figure 4.2
Estimated and Recommended Cybersecurity Budgets of Medium DIB Firms

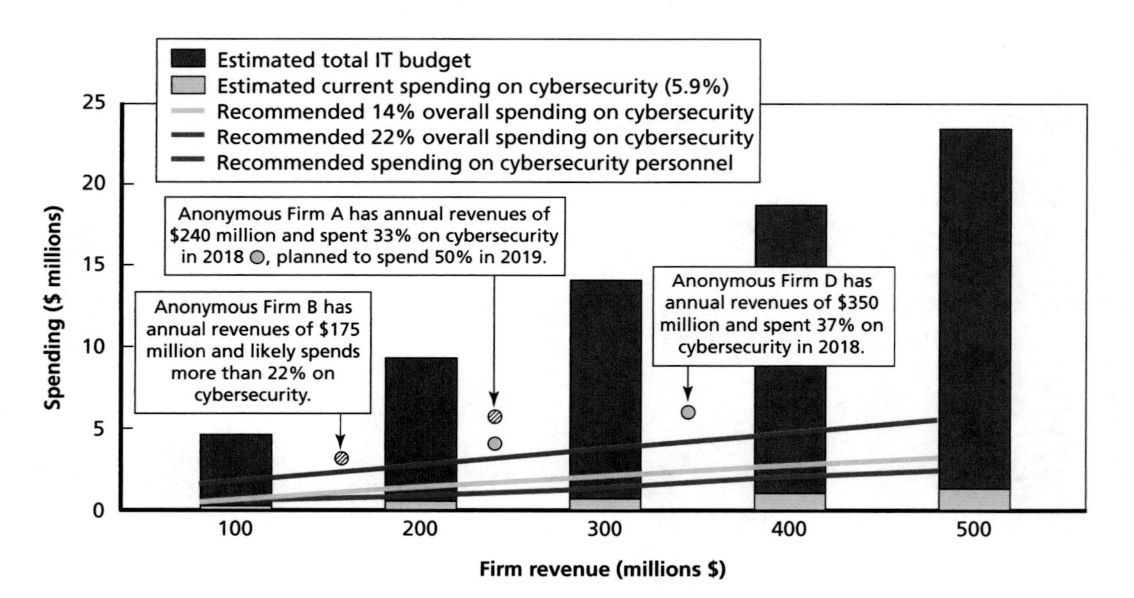

[18] Asti, 2019, p. 5.

Cybersecurity Spending by Medium-Sized Defense Industrial Base Firms

Medium-sized firms are better positioned to absorb the costs of cybersecurity tools and professionals than small firms, but they still face challenges, as shown in Figure 4.2.

As in Figure 4.1, the red line in Figure 4.2 shows the recommended number of cybersecurity professionals medium-sized firms require to secure their unclassified networks. Cybersecurity personnel costs grow as a function of the size of the firm. Based on these estimates, a firm with a revenue of $400 million will spend $1.5 million per year on cybersecurity personnel costs.

Personnel costs alone will be larger than the estimated current amount spent on cybersecurity firms of this size (lighter gray bars), unless they spend the recommended 14 percent (yellow line) or 22 percent (purple line) on cybersecurity as a percentage of their IT budgets.

Included in Figure 4.2 are estimates of the cybersecurity spending of Firms A, B, and D, we consider medium-sized. For Firms A and D, these calculations are based on the firms' responses regarding the percentage of their IT budget spent on cybersecurity; Firm B did not provide this information, and our estimate of its cybersecurity spending based on our recommended cybersecurity spending benchmarks. We calculated cybersecurity spending under the assumption that the firms' IT budgets are 4.7 percent of their revenues.[19] Note that Firm B did not feel adequate in its cybersecurity staff or budget, and so it is likely that it is spending at least 22 percent of its IT budget on cybersecurity and that the amount it estimates it needs may be higher than that and more than it can afford. The only firm that said that both its cybersecurity staffing levels and their cybersecurity budget were adequate was Firm A, which in 2020 plans to spend 50 percent of its IT budget on cybersecurity.

Overall, Figure 4.2 suggests that complying with DFARS 252 will be challenging even for medium-sized DIB firms.

Implications for the Defense Industrial Base

In this chapter, we developed estimates of how much money DIB firms of various sizes should be spending on cybersecurity (both CSTs and cybersecurity professionals). It is important to point out that more funding may not necessarily lead to more cybersecurity. Effective management, cybersecurity best practices, and employee training are also required to protect the unclassified networks of a DIB firm. Nevertheless, it will seriously hamper firms' cybersecurity efforts if they do not have enough money to spend on CSTs and cybersecurity professionals.

---

[19] We believe this is a conservative estimate. One firm shared explicit numbers regarding its IT budget, which is approximately 10 percent of its revenue. That percentage of revenue dedicated to IT is significantly greater than the averages of nearly all industry sectors and firms writ large.

Our analysis shows that most small, and even some medium-sized, DIB firms are likely not spending enough of their budgets on cybersecurity, which can result in vulnerable unclassified networks and pose legitimate national security concerns. A firm's cybersecurity budget must be large enough to pay for the salaries and benefits of cybersecurity professionals, software licensing fees for CSTs used, and the additional funding that may be required to respond to cybersecurity incidents, such as additional tools or services and advice from outside experts or companies. Ultimately, for small DIB firms, the cost of cybersecurity professionals will lead to a decision to either hire a staff of cybersecurity professionals or purchase additional CSTs. Many medium-sized DIB firms will have to make similar decisions. With such a choice, the firms' cybersecurity is at risk of being compromised. To fully protect the unclassified networks of DIB firms, alternative solutions are needed.

CHAPTER FIVE

Cybersecurity Tools

Commercial firms use a variety of CSTs to secure their networks. Larger firms typically employ a larger set of CSTs with more robust capabilities, whereas smaller firms are more likely to employ fewer tools with less capabilities. In this chapter, we review the CSTs used by small and large DIB firms to protect their unclassified networks.

Cybersecurity Tools Typically Used by Small Defense Industrial Base Firms

Figure 5.1 illustrates the unclassified network and CSTs typically used by a small DIB firm. The figure illustrates the type of devices connected to such a network, including database servers, printers, and endpoints (e.g., user computers, routers, laptops, and data storage devices). Removable media may also be attached to the network and used to download data from endpoints or servers, making the unchecked use of removable media a major security concern. In some networks, removable media use is not allowed, while in others, CSTs are used to monitor and control the use of removable media. We assume that most small DIB firms would not have such tools in place.

Network Access Control

The DIB firm network will also likely have a wireless component that enables laptops and other mobile devices to connect to the network from different parts of the firm's facilities. Wireless network access may provide attackers an easy entry point into the larger network. Wireless routers may contain vulnerabilities that can be exploited, they may use older and easier-to-crack encryption algorithms, and—in the worst case—they may not encrypt network traffic at all. In addition, wireless network routers may admit any mobile device access to the network with a stolen username and password.[1]

Not shown in the figure are the access control systems for the core network, such as Lightweight Directory Access Protocol or Microsoft Active Directory servers,

[1] It is possible to use consumer grade wireless routers to set up wireless networks that limit network access to devices with specific MAC addresses in order to prevent unauthorized devices from accessing the network.

Figure 5.1
Network and Cybersecurity Tools Used by a Small Defense Industrial Base Firm

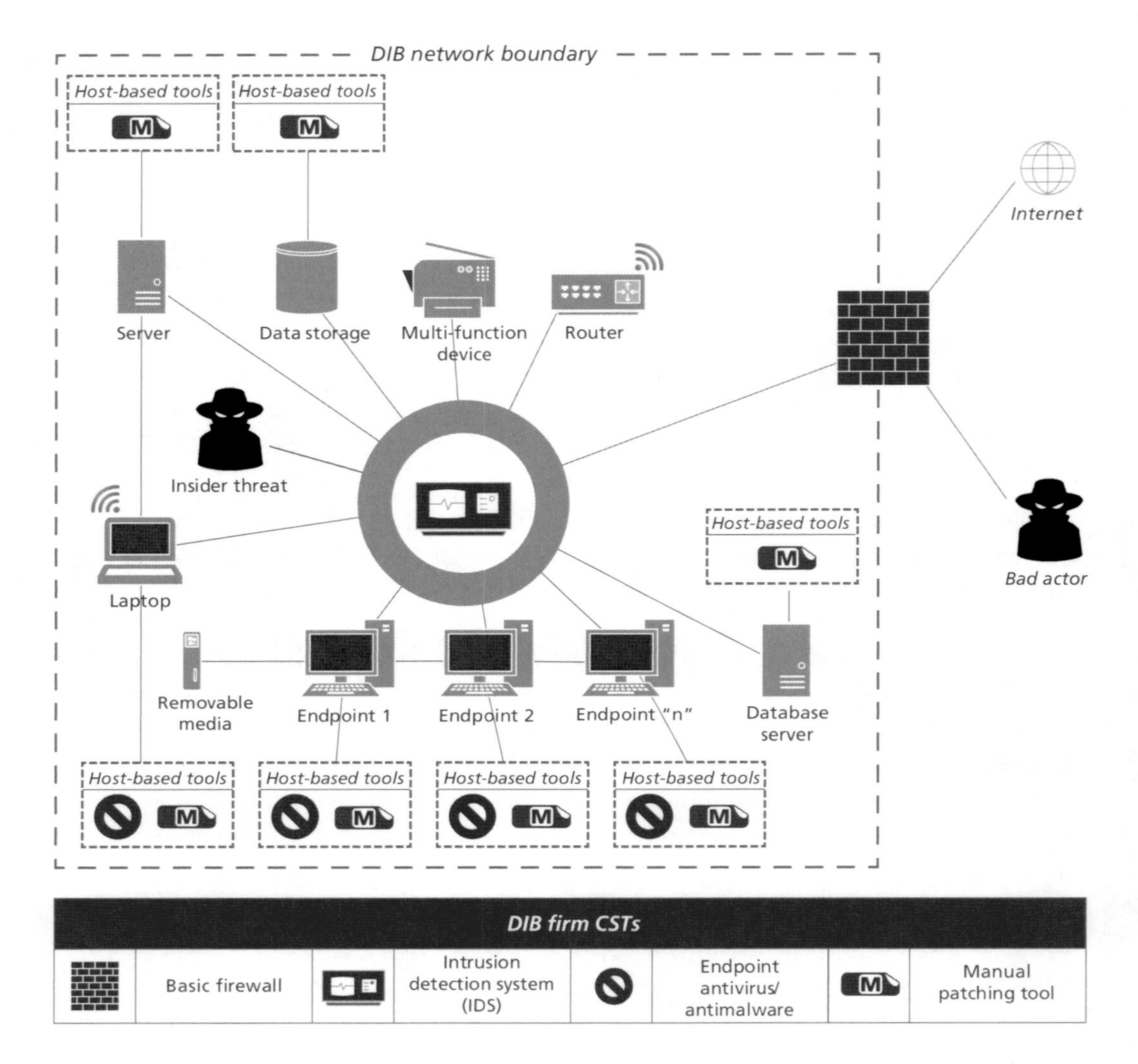

which grant or deny access to network resources. Sixty-two percent of small firms use a single-factor authentication (1FA) method for access control to their networks.[2] In most cases, this single-access control is just a username and password.[3] Medium-sized firms are more likely to use MFA methods to grant access to resources of their core networks. These factors include Internet Protocol (IP) address, device Media Access

[2] Jeff Goldman, "Most Small to Mid-Sized Organizations Don't Use Multi-Factor Authentication," *eSecurity Planet,* August 16, 2018.

[3] As noted in Chapter Three, NIST SP 800-171 requires the use of 2FA access controls. This is one reason why not a single DIB firm has yet to comply with all security controls specified in NIST SP 800-171.

Control (MAC) address, and username and password. However, traditional 1FA and MFA methods have been unable to prevent sophisticated adversaries or APTs from gaining access to network resources of DIB firms.

NIST SP 800-171 calls for at least 2FA methods, and ideally MFA methods, to be used to control access to network resources. In addition, at least one of these factors should provide a high degree of assurance that it cannot be stolen by an APT. NIST SP 800-63B describes several such methods, including hardware-based cryptographic tokens that communicate securely with an access control server.[4] We estimate that relatively few small DIB firms currently use MFA access controls that include cryptographic hardware tokens, although their use is growing among larger firms after they were first adopted for internal use by Google and Facebook. Both Google and Facebook use Yubikey, although some sections of each company use other MFA technologies in addition to Yubikey.[5] DoD uses cryptographic hardware certificates to secure its internal unclassified networks. The DoD method is more expensive because it uses public key infrastructure (PKI) encryption and CACs that contain embedded PKI encryption engines. The DoD approach does not appear to have been widely adopted by many commercial firms.

How user accounts are set up is another critical aspect of the cybersecurity of the DIB firm network. If all users are given administrative or "admin access" privileges to their own computers—or even worse, to servers in the network—this presents a security risk. If such a user's account is compromised by malware, then that account can be used to gain access to sensitive data in the network. NIST guidance in NIST SP 800-171 is to limit admin privileges to network resources to the bare minimum needed for individuals to do their jobs. Only the system administrators of the network should have widespread admin privileges across servers, network routers, and endpoints.

Network Defenses

Bad actors located inside or outside of the network will try to steal sensitive data (e.g., trade secrets or IP) by gaining access to network resources. These threats are illustrated in black in Figure 5.1, and network resources are shown in blue in the figure. An insider threat will already have access to some network resources. They may have to attack or compromise endpoints or servers on the network to get access to sensitive data they desire. In contrast, an external threat will have to gain access to the network by either exploiting a vulnerability at the boundary of the network or by inducing a

---

[4] Paul A. Grassi, *Digital Identity Guidelines Authentication and Lifecycle Management*, NIST Special Publication 800-63B, Washington, D.C.: U.S. Department of Commerce, National Institute of Standards and Technology, 2017.

[5] Juan Lang, Alexei Czeskis, Dirk Balfanz, Marius Schilder, and Sampath Srinivas, "Security Keys: Practical Cryptographic Second Factors for The Modern Web," paper presented at Financial Cryptography 2016: Financial Cryptography and Data Security, Christ Church, Barbados, February 22–26, 2016; Purdue CERIAS, "Facebook: Protecting A Billion Identities Without Losing (Much) Sleep," video," September 13, 2013.

DIB firm employee to click on a link, external website, or attachment in an email that can deliver malware to the user's machine, which can then be leveraged by the external threat to gain entrance to the DIB firm's network.

The network defenses of a typical small DIB firm are likely to be perimeter-based, with an emphasis on keeping bad actors out of the network by using firewalls and access control features of wireless routers mentioned earlier. The capabilities of network firewalls and other internet-facing servers—such as webservers—differ significantly. More-sophisticated firewalls can cost much more than basic firewalls. Also, as explained in the next section, the latest-generation firewalls are much more capable of detecting more advanced and evolving threats—including APTs—than traditional, less capable firewalls. Shown in Figure 5.1 is a basic or traditional, limited-capability firewall. Because small DIB firms are likely to be resource-constrained, we believe they are more likely to have limited-capability firewalls protecting their networks.

Other key elements of a perimeter defense-focused cyber defense architecture are antivirus detection and scanning applications that run on network-based intrusion detection systems (NIDSs) or on endpoint intrusion detection systems (IDSs). Traditional CSTs that use malware signatures to detect threats have been available on the market for over a decade and were once very effective in using signature files that were updated once a week or once a day. However, many of these older tools have proven to be less effective against new cyber threats, such as those that use encrypted payloads or polymorphic shell code.[6] Malware used by nation-states or APTs use these new technologies. APTs and malware used by sophisticated cyber criminals can evade traditional signature-based detection systems. Traditional antivirus scanners that run on endpoints are examples of systems that cannot reliably detect advanced malware threats.

Vulnerability Scanning and Software Patching

Two important tasks that need to be done regularly to protect a network are vulnerability scanning and software patching. Various tools are available to do vulnerability scanning. The results of such a scan reveal, at a minimum, the version numbers of the operating system and applications on endpoints, routers, and servers, as well as a list of applications running on hosts. Vulnerability scanning can detect unauthorized applications or processes that may have been inadvertently downloaded by a user or by malware running on a host. Scan results may also indicate whether a particular version of an application has one or more known vulnerabilities. Software vendors periodically issue software patches or new versions of their applications to eliminate known vulner-

---

[6] Yingo Song, Michael E. Locasto, Angelos Stavrou, Angelos D. Keromytis, and Salvatore J. Stolfo, "On the Infeasibility of Modeling Polymorphic Shellcode," *Proceedings of the 14th ACM Conference on Computer and Communications Security*, 2007.

abilities. Some vendors will even update their products automatically over the internet, but this required communications to be permitted by the firm's firewalls.

Manual Software Patching

In many cases, IT administrators of DIB firm networks will manually patch and update software running on endpoints and servers. A survey of small businesses reveals that only 38 percent of small firms update the software they use on a regular basis.[7] Running software that has known vulnerabilities and is not patched opens a firm's network to adversaries and cyber criminals. While this survey only looked at small *commercial* firms, there is no reason to believe that small *DIB* firms facing similar financial constraints would also not update their software regularly.

Security Information and Event Management (SIEM)

Figure 5.1 shows that the typical small DIB firm is not projected to have a cybersecurity operations center (CSOC) or be equipped with a SIEM system. A SIEM system is used to analyze CST events and alert data in real time to detect cyber attacks and data breaches, as well as collect, store, and aggregate log data from servers, endpoints, and CSTs. Even in a relatively small network, CSTs can log millions of alerts and events per day. If these alerts and log files were reviewed manually, it would not be possible for just a few—evenly highly trained—cybersecurity professionals to detect and understand whether an adversary had gained a foothold in their network.

SIEM systems assist cybersecurity professionals in aggregating data from multiple sources in the network, identifying anomalies, and, if necessary, taking appropriate action to limit the damage caused by a breach of the network. SIEM is less useful if a network is never breached, but it is essential if a network does become compromised; in this event, SIEM can assist the cybersecurity professionals in incident response and forensic analysis of the cyber attack.

Most small firms do not have a SIEM system. A recent survey of small commercial firms by Ernst and Young estimates that only 40 percent of small firms have a SIEM system.[8] This survey defined a small firm as one having revenue of $1 billion or less. In our categorization of DIB firms, this includes both small and medium-sized DIB firms, and even some large DIB firms. Therefore, it is likely that the percentage of small DIB firms, as defined in this study (those with annual revenues of less than $100 million), that have a SIEM system is likely to be less than 40 percent.

---

[7] Matt Mansfield, "Cyber Security Statistics: Numbers Small Businesses Need to Know," *Small Business Trends*, blog, January 3, 2017.

[8] Ernst & Young, "Is Cybersecurity About More Than protection? EY Global Information Security Survey 2018–19," October 10, 2018.

Implications for Small Defense Industrial Base Firms

Unclassified Networks of Small Defense Industrial Base Firms Are at Higher Risk

The discussion above implies that the cybersecurity architectures of small DIB firms are likely to be deficient in several key areas: user authentication, network defenses, vulnerability scanning, software patching, and SIEM capability or cyber attack response. In addition, as will be discussed in the next section, small DIB firms are likely to also lack other important CSTs that have been developed to respond to new threats and are now in use mostly by large firms who can afford these more sophisticated and expensive CSTs. Finally, it is important to highlight one key reason why SIEMs are so important: Employing a purely perimeter defense-based cybersecurity architecture is unlikely to be successful. The history of recent cyber breaches against government agencies and private firms of all types shows that it is likely that a small firm will suffer a cyber breach at one time or another. One 2015–2016 survey found that 55 percent of the small firms surveyed had experienced a cyber attack in the past 12 months.[9] A firm that can (1) recognize quickly that it is under attack and that its network has been compromised and (2) respond quickly to the attack to prevent the outbound flow or exfiltration of sensitive data will suffer less harm and financial loss. This is why having a SIEM capability is so important and why, without robust defenses, 48 percent of small firms simply go out of business after they suffer a cyber attack.[10]

Cybersecurity Tools for Large Defense Industrial Base firms

Figure 5.2 illustrates the unclassified network and CSTs typically used by a large DIB firm (one with annual revenue greater than $500 million). The figure illustrates in blue the type of devices connected to such a network, including database servers, printers, and endpoints (e.g., user computers, routers, laptops, and data storage devices). The devices connected to the network of a large DIB firm are much the same as those used in a small DIB firm, but the number of devices, their storage capacity, and their features will all be larger for the large firm.

Network Access Control

Large DIB firms, like smaller firms, will likely permit the use of removable media in many parts of their unclassified networks and will augment their unclassified network with wireless network access points simply because of the efficiency that such systems provide. However, large firms typically use CSTs to monitor and control the use of

---

9 Mansfield, 2017.

10 Mansfield, 2017.

removable media; we assume that most small- and medium-sized DIB firms would not have such tools in place.

Figure 5.2 shows in red the array of CSTs that a large firm is likely to employ. These include some of the same type of tools used by small firms. However, the tools used by a large firm will likely be more sophisticated and have more capabilities. For example, access control will likely be provided by an MFA system that includes software- or hardware-based encryption tokens or, at a minimum, include multiple network factors for use in authenticating users to the network (such as IP or MAC addresses). The wireless and remote network access portals of a large DIB firm are likely to be more secure because of the more sophisticated systems that a large DIB firm can afford to deploy. Such remote access systems will likely include a high integrity virtual private network (VPN) service and devices that display dynamic net entry codes that are implemented in tamper-resistant hardware (typically a key fob). The dynamic code is used as one factor in a 2FA authentication system, with the other factor being a traditional username and password. An example of such a system is the RSA SecurID authentication.[11]

A large firm is also more likely to subscribe to advanced cybersecurity services offered by external cybersecurity firms. Such services include commercial real-time threat intelligence, which can be used to update malware signatures in firewalls and IDSs. These IDSs will also be more advanced. The cybersecurity industry calls these advanced systems endpoint detection and response tools (EDRs), because they enable the cybersecurity tool vendor to dynamically update the endpoint systems with new threat intelligence and algorithms. EDRs use artificial intelligence and machine learning algorithms to identify malware. Older antivirus systems use a single digital signature and what was typically a single-hash result for the entire malware payload to identify malware. More-advanced EDR systems use a variety of parameters to screen data payloads received by endpoints or network firewalls. In addition, some EDRs will examine the behavior of a suspect application or file on the endpoint or in a locally deployed sandbox to identify and isolate potential malware. If the behavior of the file is deemed to be anomalous by endpoint machine learning algorithms, an alert is sent to the external cybersecurity vendor and to the cybersecurity operators within the DIB firm for action. Appendix B provides a review of some EDRs currently being offered on the market.

EDRs also can monitor files and software downloaded by the user from the internet. An EDR can be configured to prevent users from installing applications from the internet. Some EDRs also have so-called whitelisting capabilities and will enable users to install only applications that have been approved for use on the firms network or so-called whitelisted applications.

---

[11] RSA, "RSA SecurID Hardware Tokens," webpage, undated.

Figure 5.2
Network and Cybersecurity Tools Used by a Large Defense Industrial Base Firm

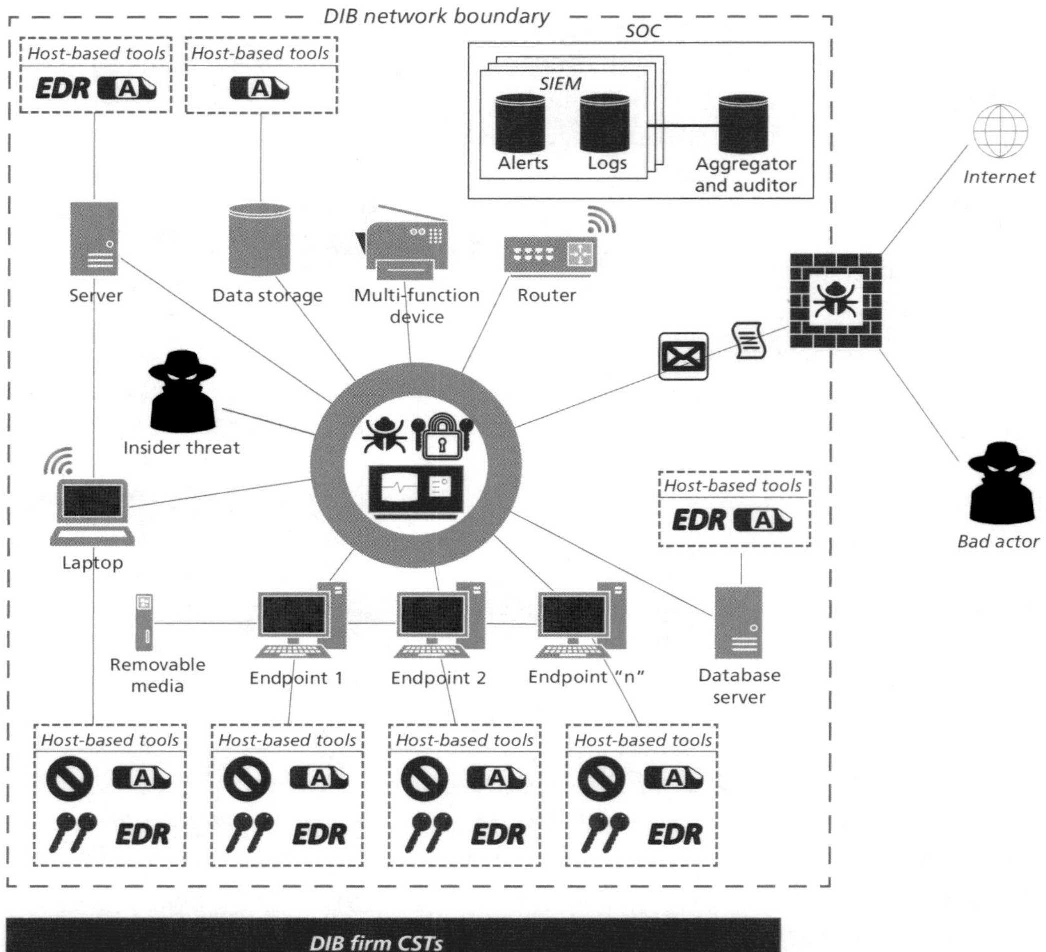

DIB firm CSTs					
	Advanced firewall		Endpoint antivirus/ antimalware	EDR	Endpoint detection and response tool
	Intrusion detection system (IDS)	A	Automated patching tool		Cyber threat intelligence
	Email security tools		Data-filtering app		Network access control (MFA)

NOTE: SOC = security operations center.

Vulnerability Scanning and Software Patching

Larger DIB firms are also more likely to use more-sophisticated vulnerability scanning and software patching tools and services. It is possible to outsource vulnerability scanning and software patching to a number of private-sector cybersecurity firms, whose services can greatly reduce the skilled labor needed to manually scan for unauthorized software, identify out-of-date software in the network, and install software patches. These tasks become more difficult and costlier as the number of endpoints in the network increases, which incentivizes larger firms to automate these processes where possible.

Email Security, Data Filtering, and Data Loss Prevention

Email provides adversaries a potential entry point into the unclassified networks of all DIB firms, regardless of size. Phishing attacks against many firm employees and executives have been successful. If the employee or executive clicks on a link or opens an attachment to an email, the host machine may be infected with malware. This attack vector has prompted many cybersecurity firms to develop email security programs that "test" attachments and web links in a local isolated sandbox environment before they are opened in order to examine their behavior to determine whether or not malware has been attached. These email security programs are frequently integrated and offered with other CSTs to provide a comprehensive defense capability or "security fabric" to protect a corporate network. Having CSTs that monitor all endpoints, servers, and network switches and routers enables the security fabric to identify links between applications and processes running in different parts of the network and to correlate these events. However, the full suite of CSTs is costlier to license than a single antivirus application installed on endpoints. We argue that more-comprehensive CST tool suites may not be affordable for small DIB firms. Appendix A provides a description of several advanced CST tool suites that are currently available on the market.

Another class of CSTs that may be used by some large DIB firms is data-filtering applications, which can be programmed to identify certain types of sensitive data to prevent such data from being emailed to external individuals or organizations, or to prevent sensitive data from being entered into an external website. Data-filtering programs vary greatly in their sophistication and features. Increasingly, some of these tools are being used to prevent financial records and PII from being sent outside of companies. Some of the simpler tools are designed for this purpose. More-sophisticated programs in this class are called DLP tools. Based on informal discussions with medium and large DIB firms, we argue that few of these firms today employ DLP tools. However, with the increased emphasis on intellectual property theft, it is possible that more DLP tools will be deployed to the unclassified networks of these firms. DLP capabilities can prevent unauthorized users from copying sensitive data and can also help protect data from destructive malware, such as ransomware. Appendix C provides a description of some of the more advanced DLP tools currently offered by cybersecurity vendors.

Security Information and Event Management (SIEM)

In the prior section, we described the importance of SIEM systems; attackers may get inside the network of even large DIB firms. In one such attack in 2011 against Lockheed Martin—the largest U.S. defense contractor at the time—cybersecurity experts of the firm used what were essentially SIEM-like capabilities to track attackers in their network and eradicate the malware from their unclassified network.[12] Figure 5.2 shows the SIEM capabilities that large DIB firms are likely to have, including the ability to aggregate data from many CSTs located throughout the network. Since the 2011 attack, SIEM technologies have advanced significantly and now include the capability to monitor cloud computing resources. Advanced SIEM systems are more flexible and adaptable, enabling them to ingest data from new sensors and use new machine learning algorithms to correlate and analyze security event data. They also can include user and entity behavior analytics, and security orchestration and automated response (SOAR) capabilities.[13]

---

[12] Kelly Jackson Higgins, "How Lockheed Martin's 'Kill Chain' Stopped SecurID Attack," *Dark Reading*, February 12, 2013.

[13] Kelly Kavanagh, Toby Bussa, and Gorka Sadowski, "Magic Quadrant for Security Information and Event Management," Gartner, December 3, 2018.

CHAPTER SIX

Alternative Defense Industrial Base Cybersecurity Protection Frameworks

In earlier chapters, we argued that small and medium-sized DIB firms lack the financial resources to acquire the CSTs necessary to secure their unclassified networks. These firms may also be unable to afford to hire and retain the cybersecurity professionals they need. We also presented empirical evidence that small firms lack many of the CSTs needed to secure their networks. Given the cybersecurity compliance requirements DoD has established for DIB firms and the limited financial resources of these firms, they will face difficult decisions regarding whether to comply or to not compete for defense contracts or defense-related work. We found, in anonymous discussions with selected DIB firms, that some have chosen not to compete for DoD work because of the estimated cost of achieving compliance with DoD cybersecurity requirements in NIST SP 800-171.

The DIB Cyber Protection Program

To address these issues, we propose a DIB Cyber Protection Program (DCP2) that will significantly improve the cybersecurity of the unclassified networks of DIB firms. We present four options for the DCP2. These options are defined by three key factors: DoD role in the DCP2, the size of the DIB firm, and where the unclassified network of the DIB firm resides.

DoD Role in the DIB Cyber Protection Program

One of the factors that defines options for the DCP2 is the role that DoD would play in the cybersecurity protection architecture. In one set of options considered, DoD would have a direct role in protecting DIB firms from cyber attacks and would be able to respond directly and in real time to an intrusion into the unclassified network of a DIB firm. In a second set of options, DoD would still manage the program but would not play a direct role in the active cyber defense of DIB firms; instead, a select number of vetted cybersecurity firms would take lead roles in proactively defending small and medium-sized DIB firms from cyber attacks.

Size of the Defense Industrial Base Firm

The second key factor is the size of the DIB firm. We place large firms in a separate category from small and medium-sized firms because we believe that large firms are more likely to have a comprehensive set of CSTs already in place on their unclassified networks and would not require the majority of capabilities offered by the DCP2. On the other hand, small and medium-sized firms may lack significant cyber protection capabilities that the DCP2 could address.

Defense Industrial Base Network Location: On Premise or in the Cloud

The third key factor that defines these options is where CUI of DIB firms is processed and stored. In one approach, small or medium DIB firms would continue to store CUI in their on-premises unclassified networks. These networks would be hardened against cyber attack by CSTs provided when the firms agree to participate in the DCP2.

The second approach we consider is to move the processing and storage components of these unclassified networks to a secure cloud computing environment. Many DIB firms are already using cloud computing services to supplement the computing capabilities they already have in their on-premises unclassified networks. The migration to the cloud is driven by the compelling cost advantages large commercial cloud service providers (CSPs) can offer. It is also important to note that CSPs also can provide a range of cybersecurity tools and services to help secure cloud computing resources. In this report, we examine DCP2 options that make use of both the economic advantages and potential cybersecurity advantages of the cloud. Commercial CSP offerings could provide a cost-effective way for DoD to rapidly enhance the cybersecurity of small DIB firms. Later in this chapter, we will propose the development of a DIB cloud as an option for the DCP2. The DIB cloud would be based on, and located inside of, one or more commercial clouds. The DIB cloud would offer standardized computer system resources (CSRs) with embedded CSTs that would be used by small and medium DIB firms. The use of the DIB cloud by DIB firms would relieve them of many cybersecurity, IT configuration control, and software patching tasks. The DIB cloud would be administered by the DCP2 program. Further details on the DIB cloud concept are discussed in specific sections below.

Participation in the DIB Cyber Protection Program Would Be Voluntary and Include Incentives

Participation in the DCP2 would be voluntary for all DIB firms. To incentivize small and medium-sized DIB firms to participate, the cost of CSTs for their unclassified network would be paid for—in whole or in part—by DoD.[1] DoD would negotiate licensing terms with vetted CST providers. Because DoD could negotiate licensing terms for

[1] The precise cost structure of the proposed DCP2 is beyond the scope of this study. The DCP2 cost structure will depend in part on CSTs licensing costs the DoD can negotiate with U.S. cybersecurity vendors.

all firms participating in the DCP2, DoD and CST vendors would benefit from economies of scale. DCP2 CST contracts would of course also be competed, with awards in each CST category going to at least two or three vendors, ensuring price competition. The tools licensed would assist DIB firms in obtaining certification to meet the cybersecurity guidelines specified in NIST SP 800-171 and related NIST publications. DoD would only pay for or subsidize the cost of CSTs used to secure the unclassified networks of DIB firms. DIB firms that decline to participate in the DCP2 would not receive any subsidized CSTs but would still have to comply with evolving DoD DFARS contract requirements (e.g., NIST SP 800-171, or with the DoD CMMC).

If the DIB firm agrees to participate in the DCP2, as discussed below, it may be necessary to move the unclassified network of the DIB firm into a DoD-approved cloud computing environment. In this cloud-based option for the DCP2, we propose that DoD provide CSRs, along with the CSTs needed to secure these CSRs. For all the DCP2 options that are cloud-based, whether with a direct or indirect DoD role, the CSTs and CSRs used by the DIB firms would be paid for in full or in part by DoD.

Some observers may be concerned that DCP2 costs could be unconstrained such that the program itself could be unaffordable. We are not suggesting that all CSRs and CSTs be paid for or be subsidized by DoD. The types and number of tools and cloud resources a DIB firm would have to pay for would be related to its gross annual revenue; for example, medium-sized firms would have to pay for a larger fraction of their CST suite than small DIB firms. The largest DoD prime contractors would not be eligible for most DCP2 CSTs, but they could still benefit in other ways, as explained below. These large DoD prime contractors would also benefit from the DCP2 participation of the smaller DIB firms in their supply chains.

A Security Operations Center SIEM Role for Small Defense Industrial Base Firms

As discussed in Chapter Five, the majority of small DIB firms do not have an internal SIEM capability and so would be challenged to respond effectively to a sophisticated cyber attack or APT intrusion. SIEM capabilities for small DIB firms that participate in the DCP2 would be provided by a centralized DIB security operations center (SOC). The DIB SOC would alert these firms in the event of intrusion into or compromise of their unclassified networks and would provide guidance on how to respond. Through participation in the DCP2, small DIB firms would be able to offload some of the most-demanding and labor-intensive functions of cybersecurity to the DIB SOC: threat analysis, data correlation, data aggregation, and alert triage.

One advantage of having a centralized DIB SOC is that it would obtain insights into attempted cyber attacks against multiple small DIB firms. This would enable the centralized DIB SOC to spot trends and develop cyber threat intelligence that could be shared across the DIB in real time.

Distribution of Controlled Unclassified Information Across Defense Industrial Base Firms

As mentioned before, some may be concerned with the potential costliness of the proposed DCP2. As such, it is important to attempt to restrain costs by applying resources where they are needed most, given certain factors. In the context of this study, one deciding factor for eligibility for the DCP2 is the level of CUI a given firm generates and has access to. CUI is not evenly distributed across all DIB firms. Some firms create and use a significant amount of CUI (e.g., trade secrets and IP). Other firms use or have access to a very limited amount of CUI, and still others never use CUI. Small DIB firms may be less likely to create or have access to CUI, but it is still possible that they could. CUI. For example, some small DIB firms may provide a critical technology or material to one or more DoD programs that cannot be sourced from anywhere else, thereby making those firms a higher priority to harden and requiring greater cyber protections.

As discussed previously, controlled technical information (CTI) is a subcategory of CUI that includes trade secrets, information about advanced technologies, and IP.[2] Some advanced technologies may provide unique capabilities for DoD platforms for which there is no substitute and enable a DoD platform to achieve a superior level of performance that adversaries cannot match. We designate this type of advanced technology and IP as high-value (HV) CUI. On the other hand, there are other advanced technologies and IP that are important for DoD programs but for which substitutes exist at higher costs and somewhat lower levels of performance. We designate this category of information as moderate-value (MV) CUI.

Small DID firms may create or have access to HV CUI. As a DIB firm's size increases, it is more likely that it will have HV CUI, as larger firms tend to play more involved roles in the research and development and production of key DoD platforms. Similarly, some medium and small firms will have no or only MV CUI. It is likely that the networks of most large DIB firms contain HV CUI because of their direct—and, in many cases, prime contractor—roles in developing and manufacturing DoD platforms and weapon systems.

To develop an affordable set of DCP2 options, we take into account these different types of CUI the firms may hold to determine eligibility for the program. Figure 6.1 provides a framework to determine the eligibility of DIB firms for the DCP2 based on firm size and the type of CUI.

Per the preceding discussion, Figure 6.2 illustrates the notional distribution of HV CUI and MV CUI across the DIB. Later in this report, we describe how small and medium DIB firms could be anonymously categorized as "HV CUI" firms, "MV CUI" firms, and firms that hold no CUI.

---

[2] National Archives, "CUI Category: Controlled Technical Information," webpage, undated.

Figure 6.1
Framework for DIB Cyber Protection Program Eligibility

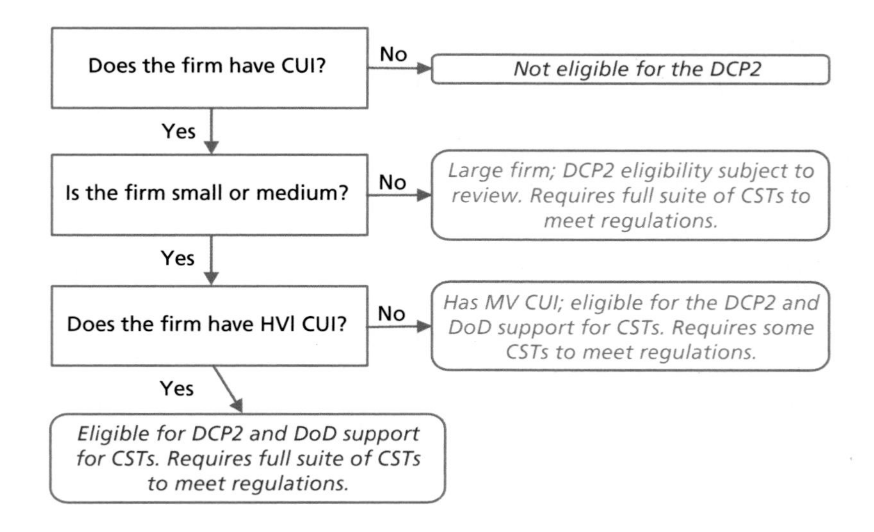

Figure 6.2
Notional Distribution of Controlled Unclassified Information Across DIB Firms

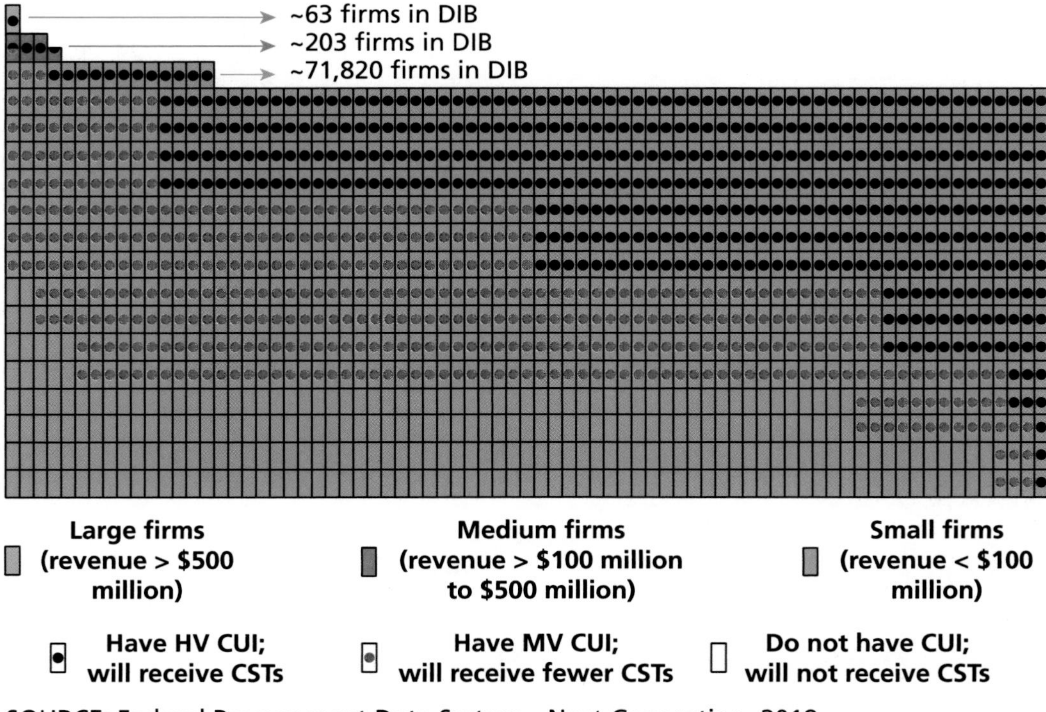

SOURCE: Federal Procurement Data System—Next Generation, 2018.

Overview of DIB Cyber Protection Program Options

Table 6.1 outlines the various categories that determine which DCP2 options will apply to a small or medium DIB firm. These categories include DoD's role (direct or indirect), the network type (on-premises or cloud-based), and the DIB firm's CUI level (HV or MV). The three categories will determine both the architecture of the option and its associated tools and services.

We also discuss the categories used to determine the various DCP2 options for large firms. As shown in Table 6.2, we do not include a CUI category, based on our earlier discussion of the likelihood that all large DIB firms create or have access to HV CUI. Therefore, we exclude a distinction between MV and HV CUI for large firms.

In the next section, we explore two options for the DCP2. The first, Option A, involves a direct DoD role for cybersecurity, and the second, Option A, involves a more indirect role. Within each option, we discuss versions in which DIB firm networks are on-premises or cloud-based. For the cloud versions, DoD will provide a DIB cloud, as described below. We discuss how each option will apply to small and medium firms versus large firms. Further, we develop two DCP2 options that apply to small or medium DIB firms based on whether they hold MV or HV CUI. The DCP2 HV CUI option will offer a full suite of tools, whereas the MV CUI option will provide a somewhat less capable and less expensive suite of CSTs. Size will also affect the distribution of tools provided by the DCP2. s

Table 6.1
DIB Cyber Protection Program Options for Small or Medium DIB Firms

DoD Role	Network Type	CUI Level
Direct role	On-premises	Moderate value
Indirect role	Cloud-based	High value

Table 6.2
DIB Cyber Protection Program Options for Large DIB Firms

DoD Role	Network Type
Direct role	On-premises
Indirect role	Cloud-based

Option A: DIB Cyber Protection Program with a Direct DoD Cybersecurity Role

In this set of options, DoD would play a direct role in defending the unclassified networks of DIB firms. DoD would establish and operate a DIB cybersecurity operations center (CSOC) that would monitor DIB firm unclassified networks.

Defense Industrial Base Firm with On-Premises Network

Here we consider how the DCP2 would apply to the unclassified network of a DIB firm located entirely inside the facilities of the DIB firm, or on premises. This is probably the option that applies to most small DIB firms today, so we consider specifically the case of a small or medium DIB firm first. In a later section, we will consider how the DCP2 would apply to a large DIB firm that had its unclassified network on premise.

We note a few primary differences between small, medium, and large firms that inform why we provide different options for each. First, small firms lack many CSTs and likely do not have an internal SOC. Second, medium firms likely have more CSTs—both in terms of number and capability—than small firms and may have an internal SOC. Lastly, large firms likely have many capable CSTs and an internal SOC. As a result of these differences, medium and large DIB firms would receive less assistance from DoD via the DCP2, but they would still benefit from participating in the DCP2 because they would receive subsidized pricing on a relevantly small number of CSTs they may be missing. In addition, they would benefit by receiving CTI, security monitoring, and alerts provided by the DIB SOC.

Small Firm with On-Premises Network and Moderate-Value Controlled Unclassified Information

As discussed earlier in this chapter, not all small and medium DIB firms may hold CUI, and if they do, its value may vary. First, we consider the case of a small or medium DIB firm with an on-premises network that has MV CUI. Figure 6.3 shows the cybersecurity architecture of such a firm that decides to participate in the DCP2. Shown in red are the CSTs the firm would continue to use, which would have been purchased and in use prior to joining the program. In Chapter Five, we surmised that a small firm would also have its own firewall and IDS. These are no longer shown in red in the figure but are shown in green instead because we assume that these small-firm CSTs will have been upgraded with advanced, more-capable firewalls and IDSs as part of their participation in the DCP2. We assume that DoD will subsidize the cost of these particular CSTs.

Several other CSTs are also shown in Figure 6.3 in green that would be provided through the DCP2, including an automated patching capability, an email screening CST, and EDRs at all endpoints. These EDRs would employ artificial intelligence (AI) to screen incoming files and monitor file behavior.

We also account for the possibility of a remote user in the network—an individual that requires remote access capability. Such a user is not shown in the figures. Typically, a VPN with 2FA allows for a remote user to securely gain access to the DIB firm network. It is important for the VPN not to be compromised or exploited by a foreign adversary and, similarly, for VPNs to be provided by a U.S. company. For example, a recent study showed that nearly one-third of VPN providers are owned by Chinese companies.[3] In support of the DCP2, we recommend that DoD use and develop an approved product list for DIB firms to avoid unintentional compromise of key information security tools by foreign actors.

The cybersecurity architecture shown in Figure 6.3 includes a comprehensive set of CSTs designed to protect the CUI resident on premise in the DIB firm's unclassified network. Some compromises have been made in the security architecture to minimize cost. These include the deployment of a simple, low-cost 2FA system for user access control and the use of data-filtering applications at key locations in the network to prevent the loss of MV CUI and also the loss of proprietary data from the contractor network to DoD. Data filtering is now offered as a standard capability on many vendor firewall offerings and EDR applications.

In this security architecture, small DIB firms with MV CUI will receive an unclassified DoD or commercial cyber threat intelligence feed to update signatures and other threat profile information in the firm's firewalls, IDSs, and EDRs.

Small Firm with On-Premises Network and High-Value Controlled Unclassified Information

Next, we consider the case of a small or medium DIB firm with an on-premises network that has HV CUI. Figure 6.4 shows the cybersecurity architecture of such a firm that decides to participate in the DCP2. Just as in Figure 6.3, shown in red are the CSTs the firm would continue to use, which would have been purchased and in use prior to joining the program. Firewalls and manual patching are no longer shown in red in the figure but are shown in green instead because we assume that these small firm CSTs will have been upgraded with advanced, more capable firewalls and IDS as part of their participation in the DCP2.

For this option, we propose a more robust cybersecurity architecture. In addition to the CSTs mentioned for small firms with MV CUI, the architecture includes a robust DLP capability that is deployed throughout the network and a cryptographically secure 2FA capability for user and network access control. We show these features in Figure 6.4.

In this security architecture, small DIB firms with HV CUI will receive an unclassified DoD or cyber threat intelligence feed to update signatures and other threat profile information in the firm's firewalls, IDSs, and EDRs.

---

[3] Jan Youngren, "Hidden VPN Owners Unveiled: 99 VPN Products Run by Just 23 Companies," VPNpro, June 2, 2019.

Figure 6.3
Option A: Cybersecurity Tools for On-Premises Network of a Small DIB Firm with Moderate-Value Controlled Unclassified Information

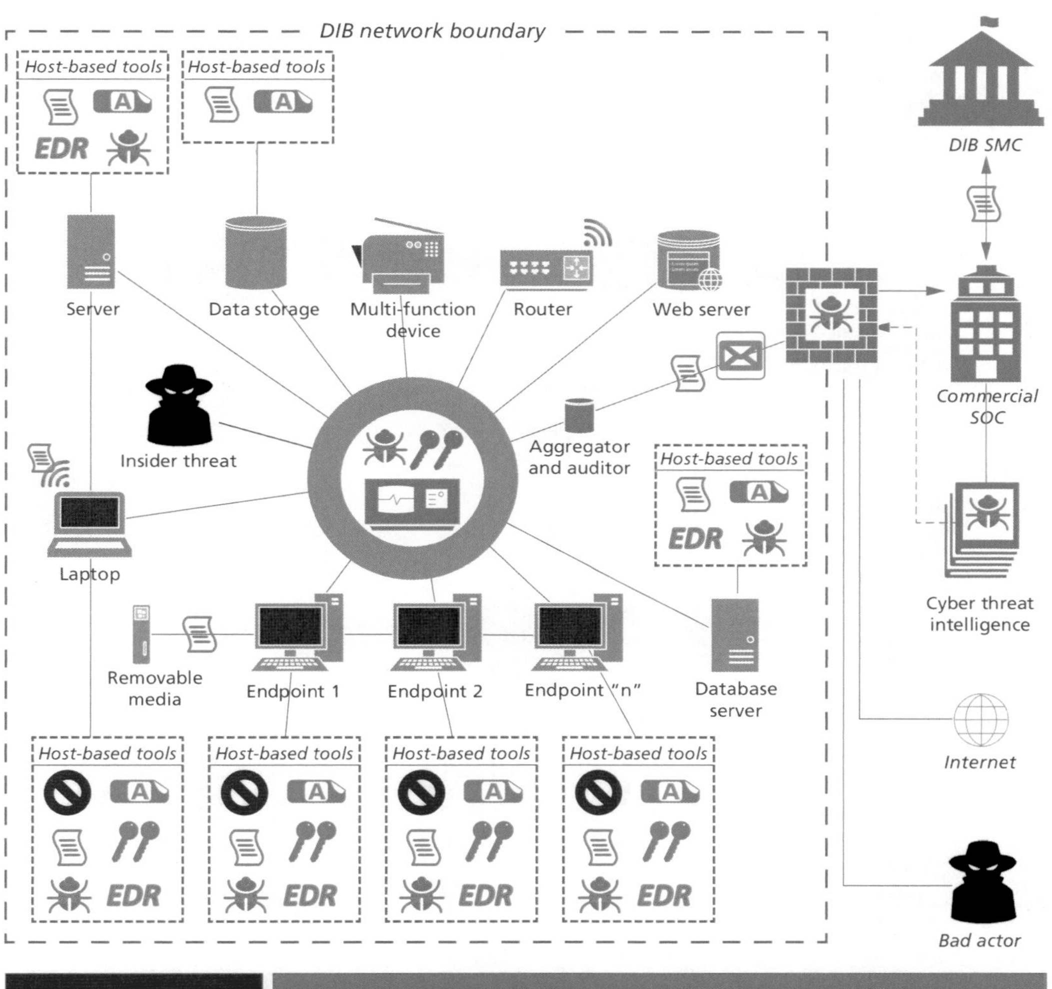

DIB firm CSTs	
(icon)	Endpoint antivirus/ antimalware

DCP2 CSTs					
(icon)	Advanced firewall	(icon)	Data-filtering app *MV CUI firms only*	EDR	Endpoint detection and response tool
(icon)	Intrusion detection system (IDS)	(icon)	Automated patching tool	(icon)	Cyber threat intelligence
(icon)	Email security tools	(icon)	Network access control (MFA) *MV CUI firms only*	(icon)	Web security tools

Figure 6.4
Option A: Cybersecurity Tools for On-Premises Network of Small DIB Firm with High-Value Controlled Unclassified Information

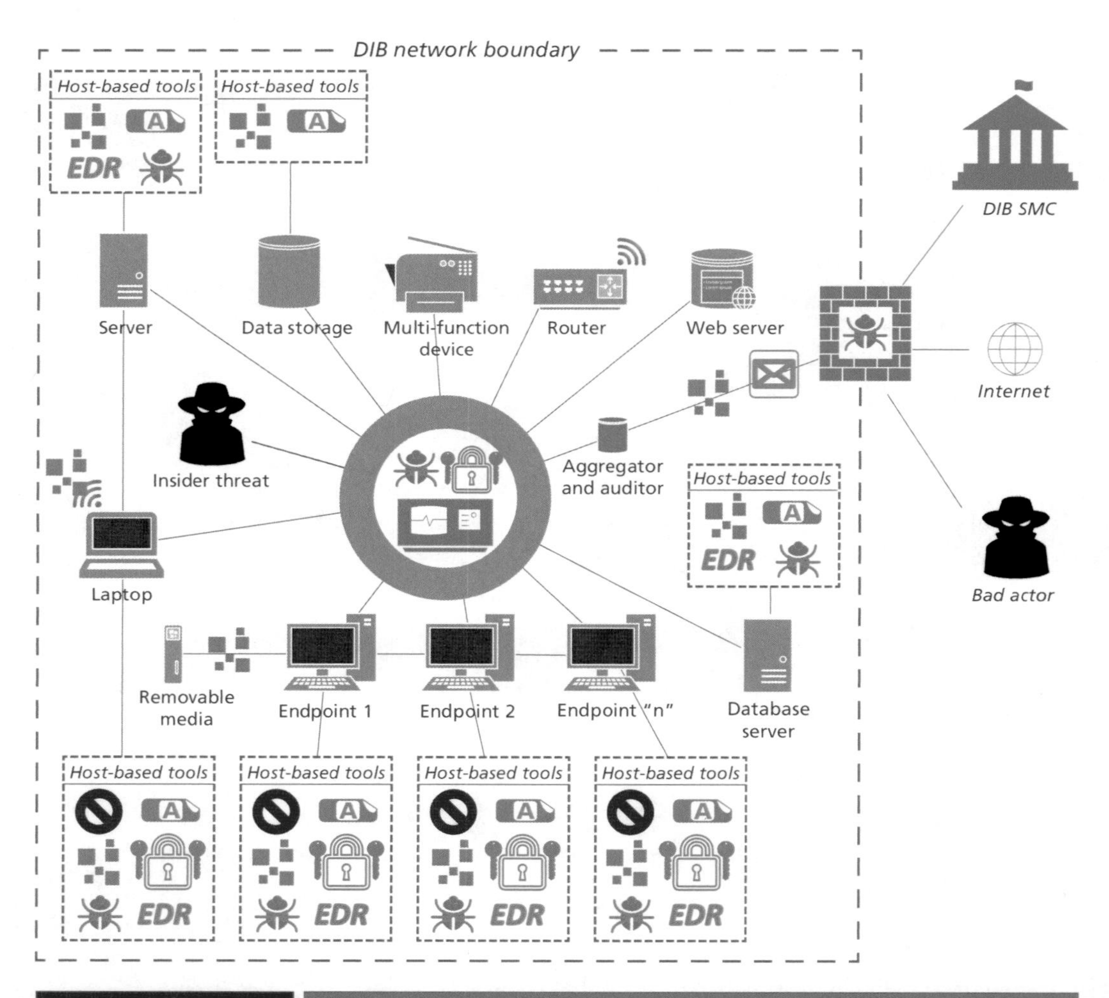

DIB firm CSTs	
(icon)	Endpoint antivirus/ antimalware

DCP2 CSTs					
(icon)	Advanced firewall	(icon)	Data loss prevention (DLP) app *MV CUI firms only*	*EDR*	Endpoint detection and response tool
(icon)	Intrusion detection system (IDS)	(icon)	Automated patching tool	(icon)	Cyber threat intelligence
(icon)	Email security tools	(icon)	Cryptographic secure MFA *MV CUI firms only*	(icon)	Web security tools

Finally, the HV CUI held by small DIB firm would also be protected by a DLP capability. The DLP capability would be used to tag or fingerprint key pieces of IP and trade secrets. If IP or trade secrets were copied by an unauthorized user, it would be detected and remediated. In addition, if an attempt was made to transmit IP or trade secrets outside the DIB network, it would be prevented. Appendix C provides a description of the capabilities of state-of-the-art DLP solutions currently being offered on the market.

Medium Firm with On-Premises Network

A medium-sized DIB firm—a firm with revenue between $100 million and $500 million—could also participate in the DCP2, and may already be operating its own SOC. If the medium-sized firm participated in the DCP2, it would send significant security events to the DIB SOC and might outsource SIEM functions to the DIB SOC to save resources. The DIB SOC, in return, would provide the medium-sized DIB firm with a cyber threat intelligence feed.

Based on our own informal discussions with medium-sized DIB firms, such firms may already have many of the CSTs shown in Figure 6.4, with the exception of EDR, MFA, and DLP capabilities. Depending on the total budget for the DCP2 program, DoD officials may see fit to subsidize the purchase of more-capable EDR, MFA, and DLP tools or to subsidize the purchase of specific CSTs missing from the networks of medium-sized DIB firms that participate in the DCP2. This would provide an incentive for medium-sized DIB firm to participate.

Large Firm with On-Premises Network

A large DIB firm—a firm with revenue of over $500 million—could also participate in the DCP2. A large DIB firm would likely already be operating its own internal SIEM and SOC. If it participated in the DCP2, the firm would still send significant security events to the DIB SOC, where they could be correlated with information from other DIB firms. Through its participation in the DCP2, the DIB SOC would provide the large DIB firm with a cyber threat intelligence feed to aid the firm's cybersecurity personnel in the defense of their network.

Based on our own informal discussions with DIB firms, a large DIB firm is likely to already have many of the CSTs shown in Figure 6.4, with the exception of cryptographically secure MFA and DLP capabilities. A large DIB firm is also very likely to possess sizable amounts of CUI. Depending on the size of the DCP2 program budget, DoD officials may see fit to subsidize the purchase of cryptographically secure MFA and DLP capabilities for large DIB firms participating in the DCP2. This would also provide an incentive for large DIB firm participation. The legal protections described later in this chapter would be incorporated into the DCP2 to encourage the participation of DIB firms of all sizes.

Network in Defense Industrial Base Cloud with Defense Industrial Base Security Operations Center

Many commercial and some DIB firms are "moving to the cloud," meaning they are moving computing and data storage resources to commercial CSPs. This migration to the cloud is occurring because of the cost savings possible when computing and application services are outsourced to market-leading CSPs such as Amazon Web Services, Microsoft, and Google. Also, moving some corporate services to the cloud can potentially increase the cybersecurity posture of a DIB firm. For example, if a firm's email servers are moved to the cloud, the CSP may be able to provide more cybersecurity resources, better software patching, and enhanced security monitoring capabilities of these cloud-based email servers than a small DIB firm could provide internally.

In this section, we describe moving the entire unclassified network of a DIB firm into a specially configured and managed DIB cloud. The options described below are distinct from the migration of some DIB firm servers and functions to the cloud. In the options considered below, CUI held by DIB firms would no longer be stored on premise. CUI would be stored and processed in a DIB cloud as described below.

Small Firm with Defense Industrial Base Cloud Network and Moderate-Value Controlled Unclassified Information

One of the potential advantages of using a cloud computing environment is that it is easier to standardize the configuration of computing resources and to maintain configuration control over a cloud-based network of computer resources. In the DIB cloud options described below, we assume that DoD, as part of the DCP2, will administer a secure enclave in a commercial cloud and develop a standardized set of CSRs for that enclave. That is, the DCP2 would provide a DIB cloud virtual machine (VM) and container repository with standardized VM and container images. The DCP2 would assume the responsibility of patching and updating this cloud infrastructure, which would be used by all participating DIB member firms. In addition, the DCP2 would establish and maintain a DIB cloud metadata service.

We envision that many vendors would compete for the CSP role for DIB cloud. One of the key metrics that should be used in selecting the one or more winning CSP contractors is the level of security they can provide. The cloud may introduce new cybersecurity vulnerabilities that have to be managed or eliminated. The DIB cloud enclave has to be set up carefully to minimize the probability of cyber intrusion. Also, the CSP must ensure that it uses a layered set of security controls in its cloud infrastructure to minimize cyber intrusions.[4] Some of these security controls appear to be consistent with those specified in NIST SP 800-171, but others are not mentioned in the draft NIST guidance document. Several CSP service offerings have been certified

---

[4] Daniel Gonzales, Jeremy M. Kaplan, Evan Saltzman, Zev Winkelman, and Dulani Woods, "Cloud-Trust: A Security Assessment Model for Infrastructure as a Service (IaaS) Clouds," *IEEE Transactions on Cloud Computing* Vol. 5, No. 3, July 2017, pp. 523–536.

to meet government cloud computing cybersecurity standards (e.g., Federal Risk and Authorization Management Program [FedRAMP] standards).[5]

DIB firms that participate in the DCP2 in this option would be provided CSRs in their own security enclaves. The security enclaves of DIB firms would be separated from one another and would provide hard security boundaries between DIB firm networks to prevent the unauthorized flow of CUI and proprietary information between potential competitors.

The proposed cybersecurity architecture and deployment of CSTs to the DIB cloud and the on-premises network of a participating DIB firm holding MV CUI are shown in Figure 6.5. The figure shows the security enclaves and virtual networks of different DIB firms in the DIB cloud. Highlighted in the figure are the CSRs of one DIB firm and its virtual network. These CSRs include data storage, servers of various kinds, and VMs.

The bottom of Figure 6.5 illustrates the on-premises network of a small DIB firm. In this cloud-based architecture, the on-premises network would consist only of thin client machines, except possibly for one or two ordinary laptops that could be used to transfer information into the network from outside sources. No CUI would be stored in the on-premises network or thin clients. All CUI would be stored and processed in the DIB cloud and would be protected by data-filtering applications running at various locations in the DIB firm cloud network enclave, as shown in the figure. The small DIB firm would be responsible for patching only the unique applications that it uses in its network, both in the cloud and on premise. The DCP2 would provide at a discounted price an automated patching capability for these applications. To manage the cost of the DCP2, a small DIB firm that only holds MV CUI would use a simple 2FA user authentication system to access on-premises and cloud resources. In addition, the DCP2 would provide only standard data-filtering capabilities at a limited number of points in the DIB firm network, as shown in the figure. Because DCP2 would provide a comprehensive set of CSTs and CSRs to the small DIB firm, the DIB firm should be able to hire more cybersecurity professionals who can operate and take advantage of the CSTs provided in the architecture.

Small Firm with Cloud Network and High-Value Controlled Unclassified Information

Figure 6.6 illustrates the unclassified network of a small DIB firm that holds HV CUI and participates in the DCP2. In this case, DoD plays a direct role in protecting DIB firms and moves the unclassified networks of participating DIB firms into the cloud. As before, the CSTs provided by or subsidized by the DCP2 are shown in green. For a small DIB firm with HV CUI, it is the same suite of CSTs shown earlier for small DIB firms participating in the DCP2 that do not use the cloud (see Figure 6.4). All servers in the DIB firm's unclassified network are hosted in the DIB cloud. Access to

[5] Federal Risk and Authorization Management Program, "Documents," webpage, undated.

Figure 6.5
Option A: Cybersecurity Tools for DIB Cloud Network of Small DIB Firm with Moderate-Value Controlled Unclassified Information

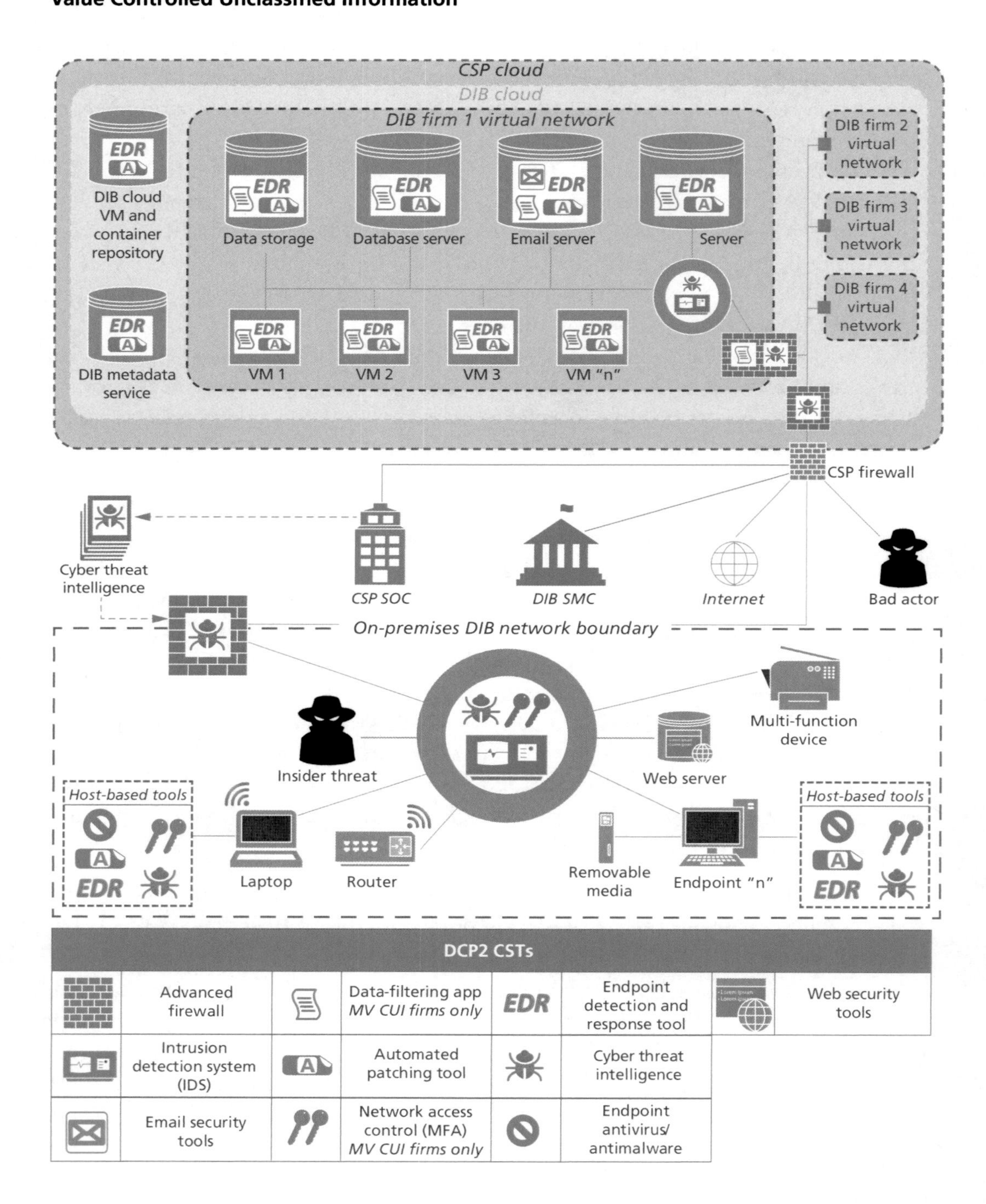

these cloud-based servers is provided through firewalls, as shown in Figure 6.6. The cloud enclave of the DIB firm is monitored by the CSP, which is presumed to operate a CSP SOC to monitor the integrity of the firm's cloud infrastructure and its firewalls. If the CSP were to detect a breach or cyber attack against one of its clients—in this case, the small DIB firm—the CSP would send an alert to the cyber operations team of the small DIB firm.[6]

The small DIB firm would also utilize the services offered by the DIB SOC. CSP alerts would be forwarded to the DIB SOC, forming a comprehensive view of the internal on-premises and cloud-based parts of the HV CUI small DIB firm's unclassified network.

The essential difference between the DIB cloud security architecture provided for a small DIB firm holding MV CUI and the HV CUI case considered in this section are the CSTs that are provided by the DCP2 to protect HV CUI. In the architecture shown in Figure 6.6, a robust DLP capability is deployed throughout the DIB cloud security enclave of the small DIB firm. In addition, the small DIB firm is provided by the DCP2 a cryptographically secure MFA capability to control access to all network resources.

Large Firm with Cloud Network

Large DIB firms are likely using CSP services for a variety of purposes, although they may not yet be storing a large portion of their IP and trade secrets in the cloud. However, even large DIB firms will likely make greater use of CSP services in the future, given the cost advantages of moving more computer and storage tasks to the cloud. As mentioned earlier in this report, large DIB firms would be invited to participate in the DCP2, and if the program decides to implement a DIB cloud, large firms could also make use of it. The cybersecurity architectures of a large firm in DIB cloud would closely resemble the architectures we have shown for small and medium firms with HV CUI (see Figure 6.6). The only difference would be the cloud resources present in their security enclave. They would use a much larger number of VMs, servers, and storage devices than small firms.

Option B: DIB Cyber Protection Program with DoD Indirect Role

In Option B, DoD would play an indirect role in the cyber defense of DIB firms. DoD would still approve and manage the distribution of CSTs to DIB firms. DoD would

---

[6] We are not presuming that the CSP would not monitor the internal communications of the DIB firm entering and exiting the cloud. We presume that the firm's message traffic will be encrypted and could only be decrypted by the DIB firm. However, the CSP would monitor the health and status of the VMs in the DIB firm cloud enclave and could potentially detect unauthorized access to those VMs and CSP-provided applications, such as email. If such unauthorized access was detected, the CSP SOC would send an alert to the DIB firm.

Figure 6.6
Option A: Cybersecurity Tools for DIB Cloud Network of Small DIB Firm with High-Value Controlled Unclassified Information

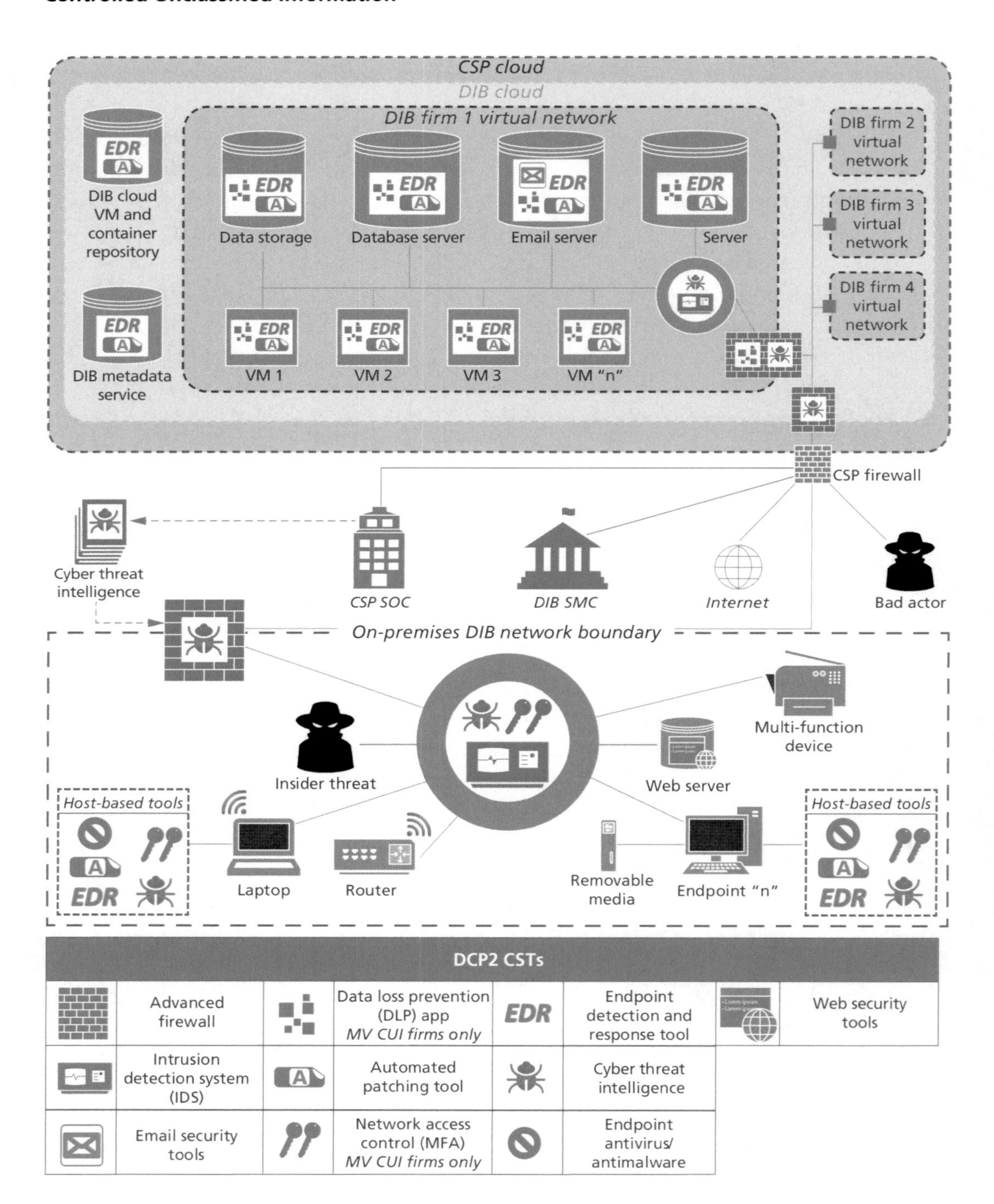

DCP2 CSTs							
	Advanced firewall		Data loss prevention (DLP) app *MV CUI firms only*	EDR	Endpoint detection and response tool		Web security tools
	Intrusion detection system (IDS)		Automated patching tool		Cyber threat intelligence		
	Email security tools		Network access control (MFA) *MV CUI firms only*		Endpoint antivirus/ antimalware		

establish a DIB security monitoring center (SMC), but it would not provide real-time SIEM capabilities to DIB firms.

Commercial Security Operations Center in Option B

In Option B, commercial cybersecurity firms would establish SOCs dedicated to the protection of DIB firms, in particular, small DIB firms. Commercial SOC contracts would be awarded on a competitive basis to leading U.S.-owned and -vetted cybersecurity firms. These cybersecurity service providers would be vetted for this role by DoD. There are several U.S. commercial cybersecurity firms that already provide this type of service (e.g., CrowdStrike and FireEye). Commercial SOCs that support the DCP2 would be located on U.S. soil, be operated by U.S. citizens, and would use only U.S. telecommunications networks to connect to DIB firms.

Network On Premise with Commercial Security Operations Center and Defense Industrial Base Security Monitoring Center

Small Firm with On-Premises Network and Moderate-Value Controlled Unclassified Information

A small DIB firm with MV CUI would be offered CSTs to participate in the DCP2, but the set of CSTs offered would be less capable and less expensive than for a small firm with HV CUI. These DCP2 CSTs are shown in green in Figure 6.7. The few CSTs purchased by the DIB firm are also shown in red.

As with small HV CUI forms, some of the DCP2 CSTs (such as firewalls and IDSs) offered to the firm may be upgraded versions of CSTs originally purchased by the DIB firm to help ensure that these CSTs meet the requirements soon to be established by the DoD CMMC. Additionally, the commercial SOCs would provide small MV CUI DIB firms with a cyber threat intelligence feed to update signatures as shown in Figure 6.8. The DCP2 would provide free of charge or subsidize the purchase of a low-cost 2FA network access-control system.

Small DIB firms with MV CUI would not be provided a DLP capability, and instead the firewalls and EDR systems provided to the firm would include a data-filtering capability to protect MV CUI.

Figure 6.7 also shows that a data-filtering capability is employed to ensure that MV CUI is not transmitted to the commercial SOC and, in turn, to the DIB SOC. The DIB SOC is not directly connected to the unclassified network of the DIB firm and would only receive sanitized CST data from the commercial SOC monitoring the DIB firm network for security threats.

Small Firm with On-Premises Network and High-Value Controlled Unclassified Information

Just as in Option A, small DIB firms with HV CUI would be offered CSTs to participate in the DCP2. These DCP2-provided CSTs are shown in green in Figure 6.8 for

Figure 6.7
Option B: Cybersecurity Tools for On-Premises Network of Small DIB with Moderate-Value Controlled Unclassified Information

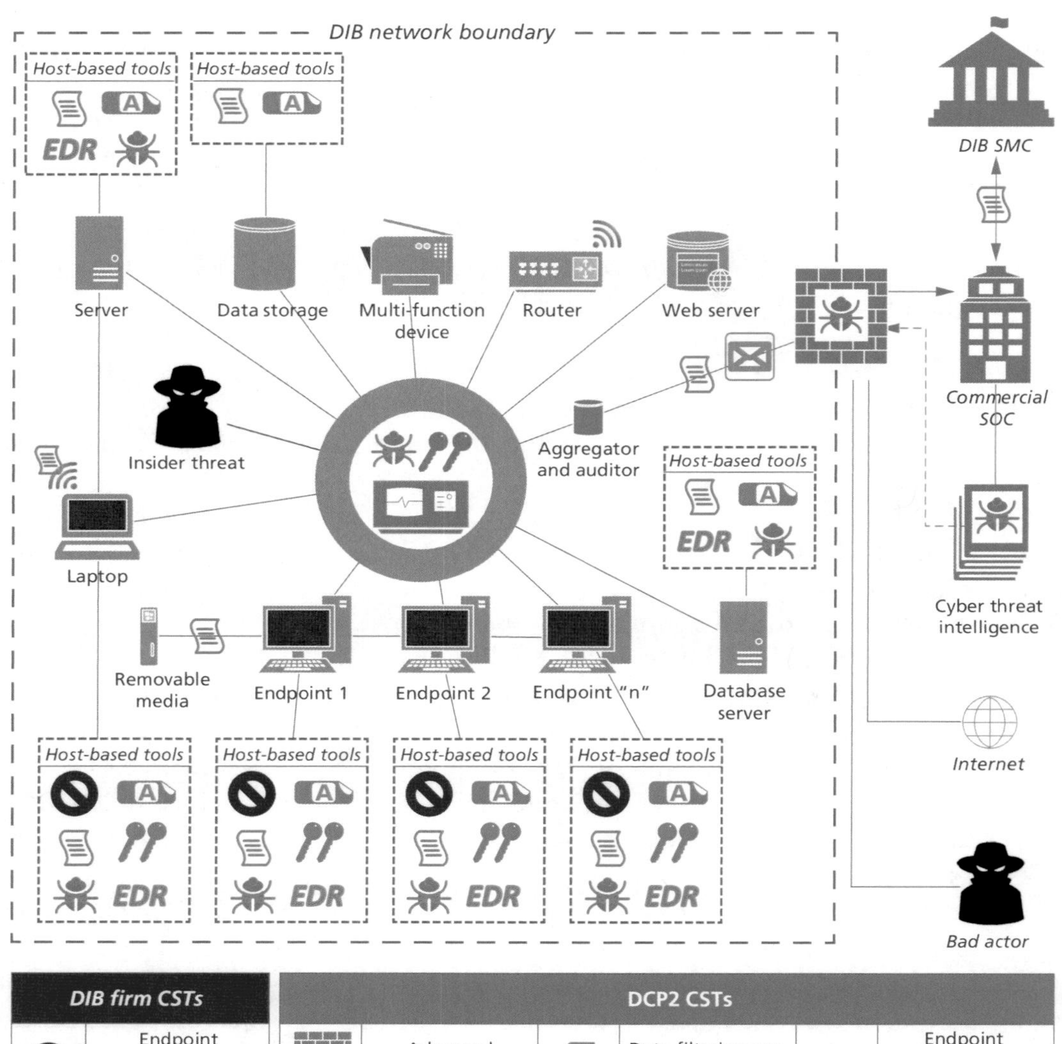

DIB firm CSTs	
(icon)	Endpoint antivirus/ antimalware

DCP2 CSTs					
(icon)	Advanced firewall	(icon)	Data-filtering app *MV CUI firms only*	EDR	Endpoint detection and response tool
(icon)	Intrusion detection system (IDS)	(icon)	Automated patching tool	(icon)	Cyber threat intelligence
(icon)	Email security tools	(icon)	Network access control (MFA) *MV CUI firms only*	(icon)	Web security tools

the case where the DIB firm's unclassified network is located entirely on premise. The few CSTs purchased by the DIB firm are shown in red.

As in Option A, some of the DCP2 CSTs (such as firewalls and IDSs) might be upgraded versions of CSTs originally purchased by the DIB firm to help ensure that these CSTs meet the requirements implied in NIST SP 800-171, or the requirements to be established in the soon-to-be-released DoD CMMC, as discussed in Chapter Three.

In Option B, the commercial SOCs would provide small DIB firms holding HV CUI with a cyber threat intelligence feed to update signatures and other threat profile information in the DIB firm's firewalls, IDSs, and EDRs, as shown in Figure 6.4. The DCP2 would provide free of charge or subsidize the purchase of a cryptographically secure MFA network access-control system for the wireless, remote, and wired parts of small HV CUI firms' unclassified DIB networks.

A small DIB firm with HV CUI would also be provided a DLP capability that could be deployed throughout its network to protect HV CUI. The DLP capability would be used to tag or fingerprint HV CUI. If HV CUI were copied—or an attempt to transmit it out of the network was made by an unauthorized user—these actions would be detected and prevented by the DLP capability. Appendix C provides a description of state-of-the-art DLP solutions currently being offered on the commercial market.

Note that the DIB SMC is not directly connected to the network of the small firm in Figure 6.4. Instead, it is connected to the commercial SOC that is monitoring the cyber artifacts and alerts generated by the CSTs in the DIB firm's unclassified network. This indirect connectivity between the DIB firm and DIB SMC would help ensure that sensitive and proprietary information of the DIB firm is not inadvertently transmitted to the DIB SMC.

Medium Firm with On-Premises Network

DIB firms of medium size (those with annual revenue between $100 million and $500 million) are likely to already be spending significant amounts of money on cybersecurity tools and staff, as we showed in Chapter Four. This means they will likely have already purchased and be using a significant number of capable CSTs. Nevertheless, based on a small set of anonymous data on medium DIB firms, we find that many of these firms still lack all the CSTs they would ideally employ to defend their unclassified networks. As mentioned earlier, we recommend that DoD encourage medium firms to participate in the DCP2. If they do participate, they would receive the CSTs needed to fill gaps in their cybersecurity architecture or, in certain cases, receive upgraded CSTs to fill those gaps. Another benefit of participation in the DCP2 for medium DIB firms is that they would receive cyber threat intelligence and intrusion alerts from the commercial SOC that would be linked to their network. As also discussed earlier in the report, it is possible that some medium DIB firms may not operate a full SOC inter-

Figure 6.8
Option B: Cybersecurity Tools for On-Premises Network of Small DIB Firm with High-Value Controlled Unclassified Information

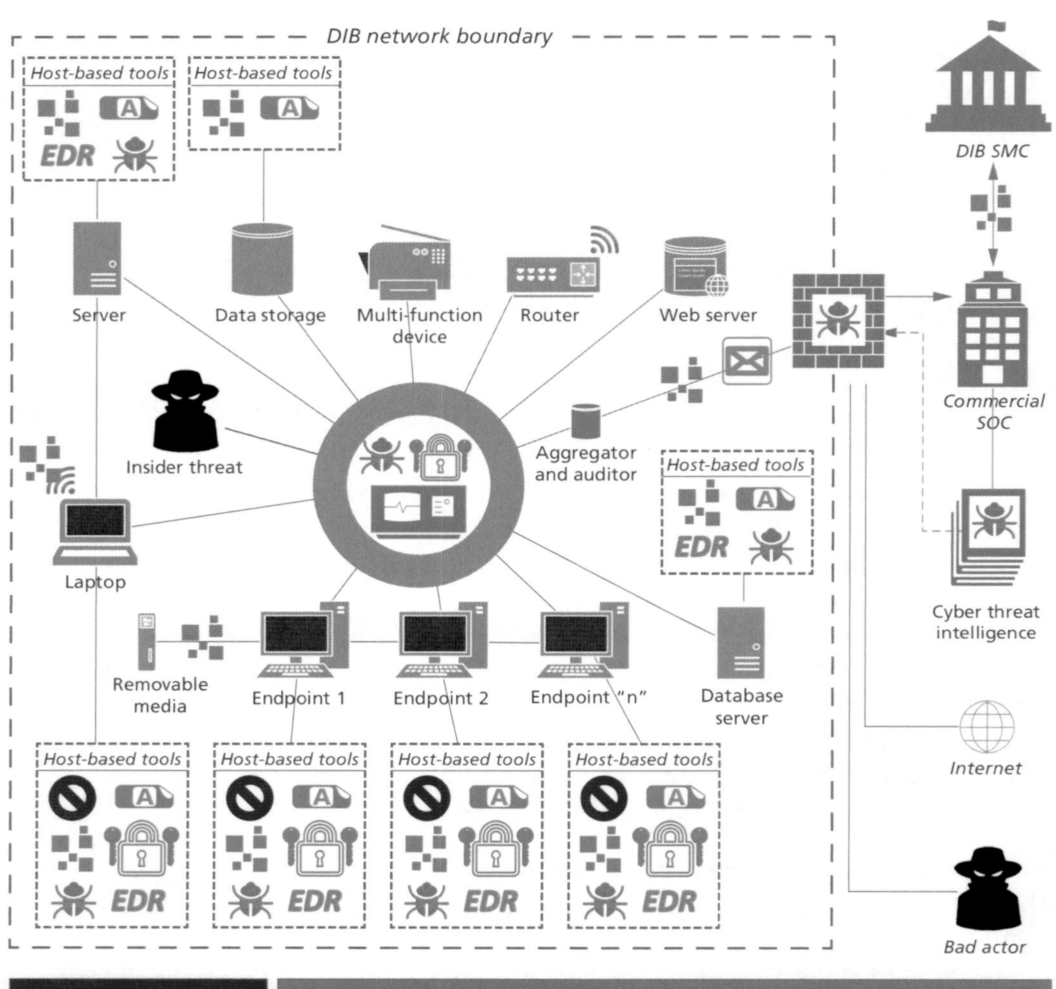

DIB firm CSTs	
	Endpoint antivirus/ antimalware

DCP2 CSTs					
	Advanced firewall		Data loss prevention (DLP) app *MV CUI firms only*	EDR	Endpoint detection and response tool
	Intrusion detection system (IDS)		Automated patching tool		Cyber threat intelligence
	Email security tools		Cryptographic secure MFA *MV CUI firms only*		Web security tools

nally. These firms would benefit from connectivity to the commercial SOC and indirectly to the DIB SMC. Even those firms that operate an internal SOC could benefit from the expertise provided by a commercial SOC and the DIB SMC.

Large Firm with On-Premises Network

DIB firms of large size (those with annual revenue greater than $500 million) are likely to already be spending significant amounts of money on cybersecurity tools and staff. They will already have a robust set of CSTs. We recommend that DoD encourage large firms to participate in the DCP2. The Option B DCP2 cybersecurity architectures for large DIB firms with on-premises unclassified networks would resemble their Option A analogs, with the following four changes:

- A commercial SOC connects directly to the DIB network and provides SIEM services.
- The DIB SOC connects only to the commercial SOC.
- DIB firms receive a cyber threat intelligence feed from the commercial SOC, and they do not receive a DoD threat intelligence feed.
- The commercial SOC forwards sanitized cyber threat and situation awareness data to the DIB SOC.

The Option B on-premises alternative for the DCP2 may be more expensive for DoD because it would require both a commercial SOC and a DIB SOC, whereas Option A only requires DoD to establish and maintain a DIB SOC. On the other hand, Option B reduces the probability that DIB firm proprietary data, including CUI not intended for release to the U.S. government, would unintentionally spillover to the DIB SOC. Option B may also provide some additional legal protections to DIB firms, as explained in the next section.

Network in Defense Industrial Base Cloud with Commercial Security Operations Center and Defense Industrial Base Security Monitoring Center

In this section, we describe moving the entire unclassified networks of DIB firms into the cloud. The options described below are distinct from the migration of a limited number of DIB firm servers to the cloud. In these DIB cloud options, CUI held by DIB firms would no longer be stored on premise. It would be stored and processed in a DIB cloud.

Small Firm with Cloud Network and Moderate-Value Controlled Unclassified Information

In this case, a CSP will provide a secure enclave in a commercial cloud and a standardized set of CSRs for that enclave for the DCP2. The DCP2 would provide a DIB cloud VM and container repository with standardized VM and container images that can be used by DIB firms. The CSP would assume responsibility for patching and updating the cloud infrastructure used by DIB member firms. In addition, the CSP would establish and maintain a DIB cloud metadata service.

As noted before, the cloud may introduce new cybersecurity vulnerabilities that have to be managed and, if possible, eliminated. The DIB cloud enclave has to be set up carefully to minimize the probability of cyber intrusion. The CSP would use security controls in its cloud infrastructure to minimize the probability of cyber intrusions.[7] Some of these security controls appear to be consistent with those specified in NIST SP 800-171, but others are not mentioned in the draft NIST guidance document. Several existing CSP service offerings have been certified to meet government cloud computing cybersecurity standards (e.g., FedRAMP standards).[8]

In Option B, the DIB firms that participate in the DCP2 would be provided a standardized set of secured CSRs in their own security enclaves, as shown in Figure 6.9. The security enclaves of individual firms would be separate from one another and would provide hard security boundaries between DIB firms to prevent the unauthorized flow of CUI and proprietary information.

The proposed cybersecurity architecture and deployment of CSTs to the DIB cloud and to the on-premises network of a participating DIB firm are shown in Figure 6.9. The bottom of the figure illustrates the on-premises network of a small DIB firm. The on-premises network would consist of thin client or thick client machines configured to prevent local storage of corporate data. No CUI would be stored in the on-premises network.

All CUI would be stored and processed in the DIB cloud and would be protected by data-filtering applications running at various locations in the DIB firm cloud network enclave, as shown in the figure. The small DIB firm would be responsible for patching only the unique applications that it uses within its network, both in the cloud and on premise. The DCP2 would provide at a discounted price an automated patching capability for these applications. To manage the cost of the DCP2, a small DIB firm that only holds MV CUI would use a simple 2FA user authentication system to access on-premises and cloud resources. In addition, DCP2 would provide only standard data-filtering capabilities at a limited number of points in the DIB firm network, as shown in the figure. Because DCP2 would provide a comprehensive set of CSTs and CSRs to the small DIB firm, the firm should be able to hire more cybersecurity professionals who can operate and take advantage of the CSTs provided in the architecture.

Small Firm with Cloud Network and High-Value Controlled Unclassified Information

In this case, a CSP will provide a secure enclave in a commercial cloud and a standardized set of secured CSRs to the DIB firm holding HV CUI. The CSP would assume responsibility for patching and updating the cloud infrastructure used by DIB firms

---

[7] Gonzales et al., 2017.

[8] Federal Risk and Authorization Management Program, undated.

Figure 6.9
Option B: Cybersecurity Tools for DIB Cloud Network of Small DIB Firm with Moderate-Value Controlled Unclassified Information

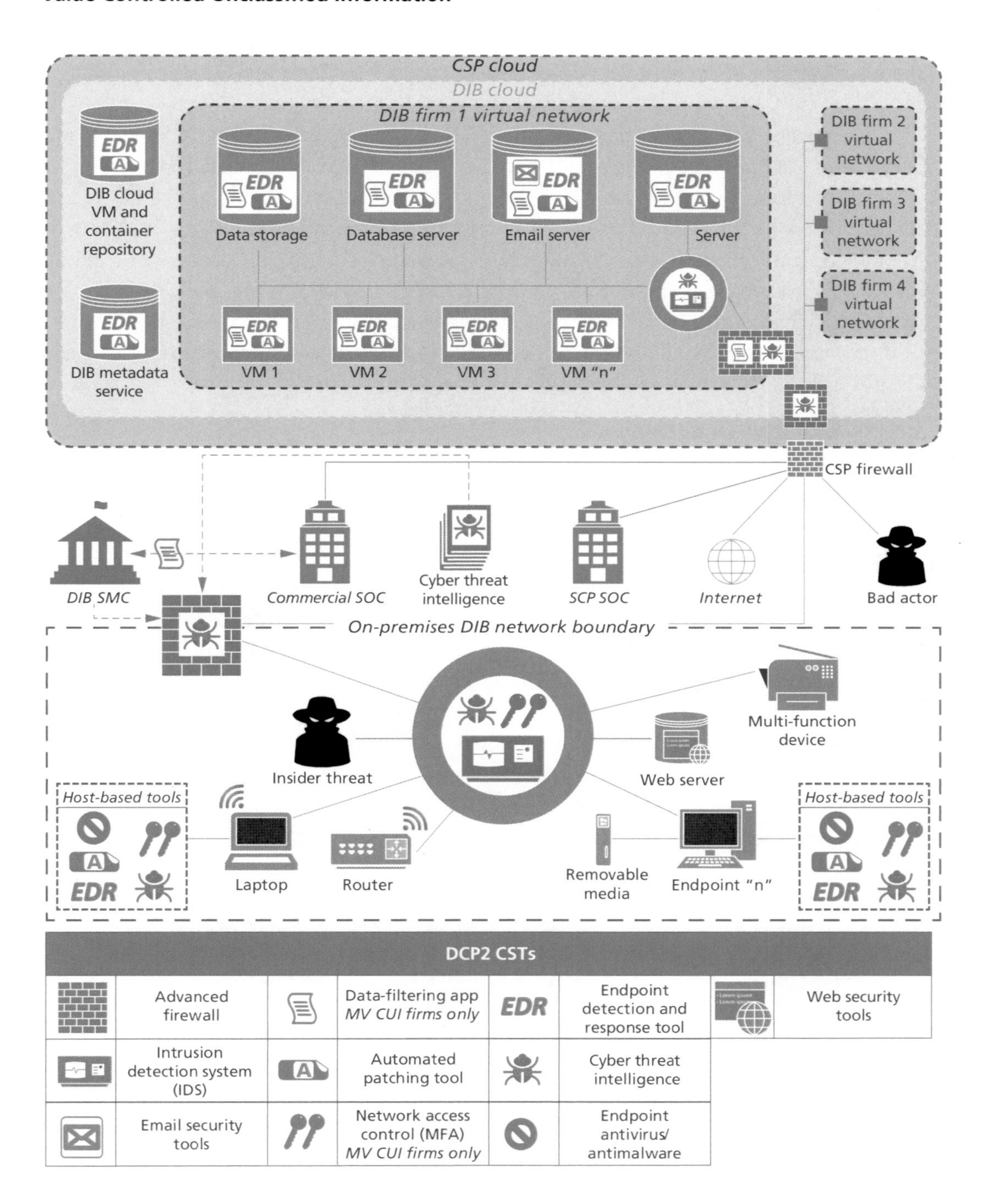

that hold HV CUI, as shown in Figure 6.10. The array of CSTs deployed to the DIB cloud network of the DIB firm and the on-premises network of the DIB firm are shown in Figure 6.10. Just as in Figure 6.9, the bottom of the figure illustrates the on-premises network of the small DIB firm. The on-premises network would consist of thin client or thick client machines configured to prevent local storage of corporate data. No HV CUI would be stored in the on-premises network.

The major differences between the DCP2 cybersecurity architecture for a small DIB firm holding MV or HV CUI are the types of CSTs provided to protect assets in the DIB cloud and in the on-premises network. As mentioned earlier, firms holding only MV CUI would be provided network equipment and EDR applications that include common data-filtering applications, as well as simple 2FA access-control capabilities. DIB firms that hold HV CUI would be provided more-sophisticated DLP capabilities that would be utilized to prevent the unauthorized data exfiltration of HV CUI and cryptographically secure 2FA access-control systems, as shown in Figure 6.10. The DCP2 would provide these CSTs at a discounted price.

Large Firm with Cloud Network

In this case, under Option B the unclassified network of a large DIB firm would closely resemble that of a small DIB firm with HV CUI. As discussed earlier in this report, a large DIB firm is very likely to hold HV CUI important for multiple DoD programs and for DoD policy.

Just as in Option A, the supporting CSP will provide a secure enclave in a commercial cloud and a standardized set of secured CSRs to the large DIB firm. The CSP would assume responsibility for patching and updating the DIB cloud infrastructure. Just as in Option A, the unclassified network of the large DIB firm would have DIB cloud and on-premises components. The array of CSTs deployed to the network of the large DIB firm would closely align with the network shown in Figure 6.10. Just as in Option A, the on-premises network would consist of thin client or thick client machines configured to prevent local storage of corporate data. No HV CUI would be stored in the on-premises network of the large DIB firm.

Large DIB firms that participate in the DPC2 would be expected to have or would be provided more-sophisticated DLP capabilities and cryptographically secure 2FA access control systems, as shown in Figure 6.10. Because of the greater financial resources of a large DIB firm, the firm may be asked to deploy these CSTs, or, if the DCP2 is provided sufficient financial resources in its budget, it may be able to subsidize the cost of these more expensive and sophisticated CSTs for large DIB firms.

Implications for Commercial Firms with Limited DoD Contracts

Not all commercial DIB firms are exclusively DoD contractors. For example, a firm may obtain over 80 percent of its revenue from non-DoD customers and only 20 per-

Figure 6.10
Option B: Cybersecurity Tools for DIB Cloud Network of Small DIB Firm with High-Value Controlled Unclassified Information

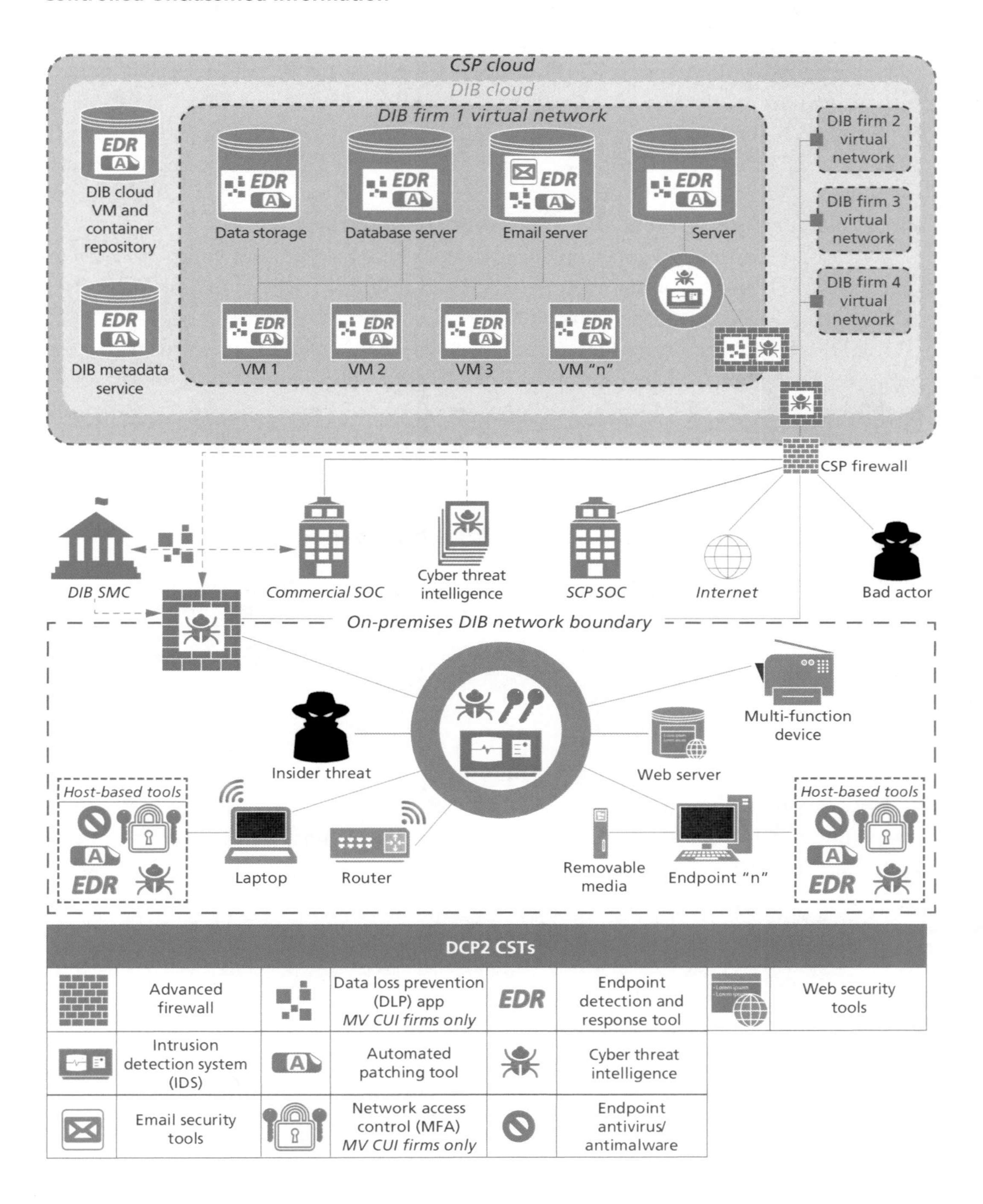

cent from DoD. In this case, it may not make financial sense for DoD to provide CSTs or cloud services for the firm's entire network. As a result, it is important to consider DCP2 infrastructure options for these types of firms. In these cases, it may also be important to consider how to manage access to firm CUI and, potentially, how to compartmentalize or separate DoD projects from other commercial activities of the firm. There are many ways in which DoD cybersecurity requirements could be met by making changes to portions of these DIB firm networks. One approach is to segment or divide the DIB firm network in DoD and non-DoD components to prevent the flow of CUI between the two network segments, and to protect the CUI associated with DoD contracts in the DoD subnetwork of the DIB firm.

In earlier sections, we described options for securing the networks of DIB firms that have CUI (these options include whether DoD has a direct role or indirect role in protecting the entire network of the DIB firm, whether the firm holds MV or HV CUI, and whether the entire network of the DIB firm is on premise or placed into the proposed DIB cloud). These options would still apply, but only to the network segment established to protect DoD-related CUI. The non-DoD segment of the DIB firm network would not be entitled to DCP2 CSTs and would remain unchanged.

Proposed Legal Issues and Protections for Both Options

In this section, we lay out potential legal issues and protections associated with both Options and B, given the service DoD would provide to DIB firms.

Again, in Option A, DoD would *directly* provide cybersecurity capabilities to DIB firms; in Option B, DoD would provide cybersecurity capabilities *indirectly*. In Option A (DoD direct role), a DIB SOC would be established to focus exclusively on protecting DIB firms. In Option B (DoD indirect role), the DIB SOC would be distributed among several leading U.S.-owned cybersecurity firms that would provide tools and services. For example, several U.S. cybersecurity companies could be selected via a competitive bidding process to operate a set of DIB SOCs that would exchange information with each other to build a comprehensive dynamic threat intelligence picture. Some combination of these two approaches also would be possible. Participation would be voluntary for all DIB firms. Either option would require DoD to spend additional resources to protect the DIB. For both options, DoD would pay for or subsidize the cost of advanced cybersecurity tools and services to secure DIB unclassified networks. In some cases, the cybersecurity services would be provided by DoD personnel located at the DIB SOC (i.e., Option A).

A DIB Cyber Protection Program would require DIB firms that wish to participate to enter into an agreement with DoD. This agreement would have the following characteristics:

- Commercial CSTs will be provided by DoD to participating DIB firms for their unclassified networks at a discount or at no cost.
- DIB firms will agree to provide sanitized data, network metadata, and cyber threat alerts to the commercial or DIB SOC.
- DIB firms will agree to take specific steps to secure CUI in their possession and their unclassified networks.
- The commercial SOC will agree to provide threat intelligence and cybersecurity threat alerts to DIB firms for free or at a reduced cost.
- The DIB SOC will agree to provide threat intelligence and cybersecurity threat alerts to DIB firms.

The legal protections for DIB firms that participate in the DIB Cyber Protection Program would derive from and be modeled on the protections established by CISA for nonfederal entities that report cyber threats or defensive measures to the federal government. In addition to the protections outlined in CISA, the DCP2 would include the following safeguards:

- Clear guidelines for the collection, sharing, and retention of data by private entities and the federal government.
- Clear guidelines for DoD use of data obtained by private entities.
- An external audit function that would allow DIB firms to ensure that their proprietary data had been segregated in the commercial or DIB SOC and remained unshared with other DIB firms.
- Data-destruction policies that would be subject to external audit.
- Clear-cut procedures for the handling of any data breach, including data breach notifications.
- Data submitted to the commercial or DIB SOC could not be used in civil or criminal litigation by any agency of the state or federal government or by other nonfederal entities.
- Privacy protections (in addition to removal of PII from cyber threat data) that protect individual privacy rights and civil liberties.
- A mechanism to address and correct wrongfully disclosed personal information.
- A private right of action for violation of data protection procedures outlined in the DCP2 agreement if the commercial SOC is maintained by a nonfederal entity or entities.
- An advisory committee—composed of at least 12 nonfederal employees who are nationally recognized experts in data privacy, data protection, cybersecurity, and electronic surveillance—would monitor and review the activities of the commercial or DIB SOC. The advisory committee would provide an annual report to the Secretary of Defense that addresses cybersecurity threats, data security, and any concerns about use of the commercial or DIB SOC for surveillance purposes.

The language of Section 104 (c) (1) of CISA would encompass the sharing of cybersecurity information by DIB firms with DoD, a federal entity. Specifically, Section 104 (c) (1) provides that

> notwithstanding any other provision of law, a non-Federal entity may, for a cybersecurity purpose and consistent with the protection of classified information, share with, or receive from, any other non-Federal entity or the Federal Government a cyber threat indicator or defensive measure.

Therefore, the sharing of cybersecurity information by nonfederal entities, such as DIB firms, with DoD via a commercial or DIB SOC is authorized "notwithstanding any other provision of law."[9] For that reason, the legal protections for nonfederal entities, such as DIB firms, that would share cybersecurity information with DoD should be aligned with the protections set out in CISA. The legal protections provided by CISA for the sharing of cybersecurity information by nonfederal entities with the federal government include the following:

- No liability for cyber threat monitoring activities on a company's own information system, or the information systems of customers, if appropriate consent and authorization is obtained.[10]
- No civil liability for information sharing with the federal government concerning cyber threats or defensive measures if the sharing is conducted in accordance with CISA.[11]
- No waiver of any applicable privileges or protections as a result of information sharing pursuant to CISA. The threat or defensive measures information shared will not be subject to state or federal Freedom of Information Act laws; the information can be designated as the proprietary information of the private company that shares it.[12]
- No antitrust liability for two or more companies that share threat information or provide cybersecurity assistance. "It shall not be considered a violation of any

[9] "Consistent with CISA, non-federal entities may also share cyber threat indicators and defensive measures with federal entities through means other than the Federal government's capability and process operated by DHS" (U.S. Department of Homeland Security and U.S. Department of Justice, 2016, p. 15). Section 103(a) and Section 103(a)(3) of CISA state that the Secretary of Defense, among other federal entities, "shall jointly develop and issue procedures to facilitate and promote . . . the timely sharing with relevant Federal entities and non-Federal entities, or the public if appropriate, of unclassified, including controlled unclassified, cyber threat indicators and defensive measures in the possession of the Federal Government."

[10] CISA Section 104(a)(1) and Section 106(a).

[11] CISA Section 106(b).

[12] CISA Section 105(d)(1)

provision of antitrust laws for private companies to exchange or provide a cyber threat indicator or defensive measure."[13]

- No duty is imposed on private companies to share cybersecurity information, and there is no requirement that a company act on the receipt of cyber threat information. A federal entity may not "condition the award of any grant, contract, or purchase" on the reciprocal exchange of cyber threat information. [14]
- No new government regulations may be created from cyber threat or defensive measures information provided pursuant to CISA except regulations concerning how information systems can prevent or mitigate cybersecurity threats.[15]

The safeguards outlined above for the commercial or DIB SOC, combined with the legal protections provided by CISA for the sharing of cybersecurity information with the federal government, should provide incentives for DIB firms to participate in DPC2.

---

13 CISA Section 104(e).

14 CISA Sections 108(i) and (h).

15 CISA Section 105(d)(5)(D)(i).

CHAPTER SEVEN

Conclusions

Cyber attacks against DIB firms designed to steal sensitive data, trade secrets, and IP are growing in sophistication and severity. Attacks on unclassified networks in particular have increased. The current DoD approach to prevent these attacks from being successful is based on DFARS 252.204-7012 and NIST SP 800-171 and appears to be inadequate. As of July 2019, no DIB firms have been certified by an external third party to have fully implemented the cybersecurity controls specified in NIST SP 800-171.[1] In addition, our cost analysis shows that small DIB firms and some medium-sized firms do not have the resources to comply with NIST SP 800-171. Furthermore, DFARS 252.204-7012 assumes that CUI flows down from the prime contractors, with primes responsible for denying a subcontractor access to CUI if they do not comply with the regulation. However, many subcontractors are in business because of their trade secrets—CUI exists at all levels of the supply chain—and this information must be protected but is overlooked in the current clause. Therefore, we conclude that if the current approach is followed, it may not be possible to protect a significant amount of CUI held by DIB firms on unclassified networks from foreign adversaries. The persistent attacks and hemorrhaging of critical information and technology from unclassified networks, coupled with the significant financial losses, erode the U.S. DIB and threaten U.S. military advantage over the long term. This report presents an alternative approach to bolster the cyber protection of the unclassified networks of DIB firms and better position DIB firms to defend against sophisticated nation-state cyber threats.

Findings

Unclassified Networks of Small Defense Industrial Base Firms Are at Higher Risk

Our research reveals that the cybersecurity architectures of small DIB firms are likely to be deficient in several key areas: user authentication, network defenses, vulnerability scanning, software patching, and SIEM or cyber attack response. Small DIB firms also probably lack other important CSTs that have been developed to respond to new

[1] Sera-Brynn, 2019.

threats because they cannot afford to procure and operate them. It is also important to highlight why SIEM systems are so important: Employing a purely perimeter defense-based cybersecurity architecture is unlikely to be successful. A SIEM capability is essential to detecting, isolating, and extracting malware after it has gained access to the network.

The Current DoD Approach and Proposed CMMC Process Are Likely Unaffordable for Many Small and Some Medium-Sized Defense Industrial Base Firms

The current DoD approach has been criticized by industry as being unaffordable. Some industry observers have stated that it would deter small firms from bidding on DoD contracts, as it may not be feasible for small DIB firms to comply with the security control guidance issued by DoD.

The current DoD approach is also now being revised to include a cyber maturity model and a mandatory certification process, the CMMC, but it is still a compliance-based and will impose costs on DIB firms. Our cost analysis indicates most small DIB firms may not be able to afford the cyber defenses that could be mandated by the CMMC. Many medium-sized DIB firms may face the same challenges, especially if they are held to the highest compliance levels of the CMMC.

Voluntary DoD Cyber Threat Sharing Service Is Not Available to Many Defense Industrial Base Firms

The voluntary program for cyber threat information sharing has its difficulties as well. Not all DIB firms can use this service. To use the website, a DIB firm user must be able to log into the site with a DoD CAC. Some defense contractors may not have anyone on staff with these credentials. In addition, other even more valuable data are shared with DIB firms using other means and personal connections. Some DIB firms may lack the informal ties to the Intelligence Community that would make them privy to this other cyber threat information.

New Cybersecurity Tools Can Significantly Strengthen the Cyber Defense of Defense Industrial Base Firms, but They Cost Money

Cybersecurity firms have made security advances and now offer new CSTs that can help to strengthen the cybersecurity of the DIB, but these new tools are costly. In Appendix B, we describe EDR, cyber threat intelligence, and other advanced tools that are available from leading vendors in the industry. In Appendix C, we described the capabilities of two leading DLP vendors. Although our informal tool survey was not exhaustive, we found a growing array of highly capable CSTs from U.S.-owned and -operated cybersecurity firms that can be leveraged by DIB firms, but at a cost.

Recommendation

Proposed DIB Cyber Protection Program

To address the cybersecurity vulnerabilities of the unclassified networks of DIB firms, we propose that DoD establish a DIB Cyber Protection Program (DCP2) to protect the unclassified networks of DIB firms from cyber attacks of foreign nation-states. Participation in the DCP2 would be voluntary for DIB firms. DIB firms participating in the program would agree to install and use CSTs provided by DoD. These CSTs would be provided either free of charge or at significantly reduced licensing costs.

In turn, the DIB firm would agree to provide sanitized data produced by the CSTs to a new DIB SOC or a commercial SOC devoted exclusively to defending the DIB, to improve the real-time monitoring and health of the DIB.[2] These data would include network metadata, application metadata, anonymized user account metadata, security alerts, and anonymized system log files. This sanitized data would not include PII of DIB firm employees, proprietary firm information, employee correspondence, or any CUI. DoD would provide free of charge a data-sanitization application to ensure that only relevant cybersecurity data are transmitted to the DIB SOC or commercial SOC.

The DIB SOC or commercial SOC would agree to provide dynamic intelligence, security alerts, and recommended actions to DIB firms to identify and remediate APT incursions and to prevent the exfiltration of CUI from the unclassified network of the DIB firm. DoD would take the lead in purchasing CSTs that are not within reach of many small and some medium-sized DIB firms. Using economies of scale and market knowledge, DoD could negotiate better software licensing terms by making a bulk purchase for CSTs. In exchange for receiving these tools and services, DIB firms would agree to take steps to secure CUI on their unclassified networks.

The DCP2 would be beneficial to all DIB firms, including the largest prime contractors; DoD, and the Intelligence Community in that it would enable real-time threat intelligence to be collected and synthesized across the DIB in way that is currently not possible. The DIB SOC or commercial SOC would use these and the other data to generate alerts that would be sent back to DIB firms to secure and improve the monitoring of their networks from external and internal threats.

The DCP2 would provide disadvantaged yet important DIB firms access to CSTs in a way that incentivizes their participation and protects DIB firm CUI and the CUI in their supply chains. Similarly, the DCP2 would provide DoD real-time insight into the cyber health of the DIB and help identify and respond to cyber threats.

The DCP2 would not replace the proposed CMMC. It is designed to complement it and better position DIB firms to comply with NIST SP 800-171 guidance.

---

[2] The commercial SOC would be run by a vetted and cleared U.S. cybersecurity service provider. This service would be paid for by the DoD.

We recognize that the DCP2 would impose significant new cost on the government, a cost that some could argue should instead be borne by private industry, given that it will benefit in many ways from the CSTs provided by DoD. However, we argue that it is the U.S. government's responsibility to protect the DIB. DIB firms are under cyber attack by competent nation-states using significant resources that in many cases greatly exceed those available to DIB firms. Just as in other domains, private companies should be and are protected from criminal actors by law enforcement agencies (e.g., by local police departments or the FBI.[3] DIB firms are entitled to some form of cybersecurity protection by the U.S. government as well, although providing such protection requires a partnership across public and private entities to be successful.

DIB Cyber Protection Program Options: Moving Defense Industrial Base Unclassified Networks to a Defense Industrial Base Cloud

The most cost-effective option for implementing the DCP2 may be based on cloud computing capabilities. In this option, DoD would establish a DIB cloud that could be used by DIB firms for computing and storage of unclassified data. DIB firms would move their unclassified networks into the DIB cloud. If a DIB cloud were implemented, the CUI held by DIB firms would no longer be stored on premises. It would be stored and processed only in the DIB cloud.

The CSP would provide a secure enclave in a commercial cloud and a standardized set of CSRs for that enclave for the DCP2. The DCP2 would provide a DIB cloud VM and container repository with standardized VM and container images that can be used by DIB firms. The CSP would assume responsibility for patching and updating the cloud infrastructure used by DIB member firms. DoD would also establish and maintain a DIB cloud metadata service.

DIB firms that participate in the DCP2 would be provided a standardized set of secured CSRs in their own security enclaves. The security enclaves of individual firms would be separate from one another and would provide hard security boundaries between DIB firms to prevent the unauthorized flow of CUI and proprietary information.

The on-premises network would consist of thin client or thick client machines configured to prevent local storage of corporate data. No CUI would be stored in the on-premises network.

Legal Protections and Agreements

We propose that the DIB SOC be isolated from all other cybersecurity analysis centers in DoD, the Intelligence Community, and law enforcement agencies. If the FBI were to take the lead in protecting the unclassified networks of DIB firms, this could

[3] However, it is important to note that the DoD has lead responsibility for protecting DIB firms under U.S. law, not the FBI. If the FBI were to take the lead role in protecting DIB firms from cyber attack, this would introduce significant legal concerns and complications.

potentially expose firm employees and corporate officers to unrelated law enforcement investigations and actions, which potentially could violate their fourth amendment rights. As discussed earlier in this report, we propose that significant legal protections be offered to participating DIB firms as part of the DCP2 to minimize any chance that additional liabilities would be incurred by the DIB firm or its employees.

DIB Cyber Protection Program Options: A Direct or Indirect DoD Role in the Program

We identified two options for implementing the DCP2. In Option A, DoD would play a direct role in real-time cyber defense of DIB firms. To facilitate this, the DIB SOC would be directly connected to the unclassified networks of DIB firms. The DIB SOC would provide sanitized dynamic intelligence, alerts, and recommended responses to DIB firms and, in turn, would deliver cybersecurity data collected by CSTs to the DIB SOC.

In Option B, DoD would play an indirect role in real-time cyber defense of DIB firms, and a commercial SOC would be directly connected to the unclassified networks of DIB firms. The commercial SOC would provide dynamic intelligence, alerts, and recommended responses to DIB firms, which, in turn, would deliver CST data to the commercial SOC. The commercial SOC would also be connected to the DIB SOC, which would aggregate data from multiple commercial SOCs to monitor the health of the DIB. Option B would reduce the probability that privately owned CUI or sensitive DIB firm data would be inadvertently sent to DoD. It may also present fewer legal concerns to some DIB firms. However, it may be more expensive, as it would require more SOCs to be established and operated.

Maintaining Supply Chain Confidentiality While Increasing Transparency of the Defense Industrial Base

One of the long-standing challenges of administering a program like the proposed DCP2 is to ensure that DCP2 resources are made available to all DIB firms with CUI. Some smaller firms may not currently know they are part of a DoD supply chain. Such firms may provide a critical technology to an intermediate level contractor that wishes to hide the source of the critical technology from DoD prime contractors for competitive reasons.

A second challenge with implementing the DCP2 is maintaining the confidentiality of DIB firm supply chain relationships and product or component sources. Private companies do not want to disclose their key suppliers to outside parties, competitors, or DoD. Supply chain relationships are considered to be proprietary by many firms.[4] For DoD to authenticate that a company is eligible for CSTs and cybersecurity services

[4] This observation is based on private communications with defense contractors.

from the DCP2, it would need some way of doing so that does not require revealing the supply chain relationships of DIB firms to DCP2 government program managers.

Proposed DFARS Flow-Down Clause for Controlled Unclassified Information

For the DCP2 to be successful, it has to preserve supply chain confidentiality while fostering greater DIB transparency and verification of which firms have CUI. We propose that a new DFARS clause be included in DoD contracts that requires DIB firms to declare whether the DIB firm holds CUI and whether its immediate subcontractors hold CUI. The DIB firm would be required to declare the type of CUI it holds that is pertinent to DoD. The exact nature of the CUI does not have to be disclosed to the government, but the existence of the CUI would have to be shared. DoD would use this information to decide whether the DIB firm or any of its subcontractors are eligible for the DCP2.

This new DFARS clause would flow down to subcontractors, meaning that the contracts between the prime contractor and its subcontractor would contain this clause. The flow-down of the DFARS contract clause would require the subcontractors to disclose to the government whether they hold any CUI pertinent to DoD. This would ensure that DoD would obtain at least two CUI declarations for a subcontractor: one from the subcontractor itself and one from the DIB firm above it in the supply chain for the DoD program. In this way, DoD would be able to obtain a comprehensive list of DIB firms with CUI that should be eligible for the DCP2. DoD would use this information to grant membership to DIB firms into the DCP2. This approach preserves the confidentiality of the supply chain based on the DFARS flow-down clause, because CUI declarations would be made directly only to DoD.

Potential Next Steps

To determine whether a DIB firm is eligible for the DCP2, we proposed a DFARS contract clause that requires DIB firms to declare the CUI in their possession to DoD. A general CUI declaration may be insufficient to determine whether a particular firm should receive DCP2 benefits and whether the DIB firm holds MV or HV CUI. Additional information may be needed from DIB firms determine DCP2 benefits. Further research should be done to regarding pertinent DFARS contract language. Other sources of information should also be investigated to classify and prioritize CUI from a larger DoD perspective.

The proposed DCP2 presented in this report applies only to U.S. firms, but the supply chains of DoD programs can include foreign firms. If the DCP2 is to be comprehensive, it should include at least some foreign firms that supply key technologies to DoD programs. A key policy question is whether and how the DCP2 could be extended to protect the unclassified networks of key foreign suppliers.

If a DIB cloud is pursued, and if DoD supplies all cloud infrastructure for firms, firms will be getting more than just cybersecurity; they will be getting computing and storage resources for free. We recommend follow-on study on alternative economic models for the DIB cloud option that still provide incentives for small DIB firms to participate in the DCP2 but that are affordable for DoD.

There is still work to do to determine the cost of the proposed DCP2 to DoD for Options A and B. To be sure, the DIB SOC and commercial SOC costs should be examined in detail. In addition, CST licensing costs and models should be explored that include economies of scale and pricing options. It will be not a reasonable economic proposition to offer CSTs to every DIB firm. Thresholds and limits will have to be established to determine to the number of CSTs paid for by DoD, and different CST subsidy models should be explored.

DIB firms may be ambivalent about sharing network and application metadata and anonymized account behavior data with DoD. However, the cybersecurity industry has developed CSTs that sanitize cyber artifacts, which can be used to detect anomalous behavior without sending internal contents of files to an external SOC. Further research on CSTs is required to confirm these claims and to determine when additional data-sanitization tools will be needed to preserve the privacy and Fourth Amendment rights of DIB firms and employees.

Finally, it will be important to manage the cost of the DCP2. Only DIB firms that hold important CUI and provide DoD with critical defense-related technologies should be eligible to receive the full benefits of the program. Smaller firms that supply mostly commodity-related items to defense programs may not be eligible. A parametric cost analysis should be conducted that estimates the cost of the DCP2 that varies the number of DIB firms with important CUI.

APPENDIX A

Detailed Network Diagrams for Cyber Protection Framework

This appendix provides notional network diagrams developed to illustrate each DCP2 option discussed in Chapter Six—Option A (DoD-lead DIB SOC) and Option B (a commercial-lead SOC)—as well as diagrams for the status quo scenarios and options. These network diagrams demonstrate where each given tool or capability within the DCP2 would be deployed on the network of DIB firms of varying sizes: large DIB firms, large DIB firms with a cloud environment, small DIB firms, and small DIB firms with a cloud environment.

Status Quo

Figure A.1
Network Diagram for a Large Firm (Status Quo)

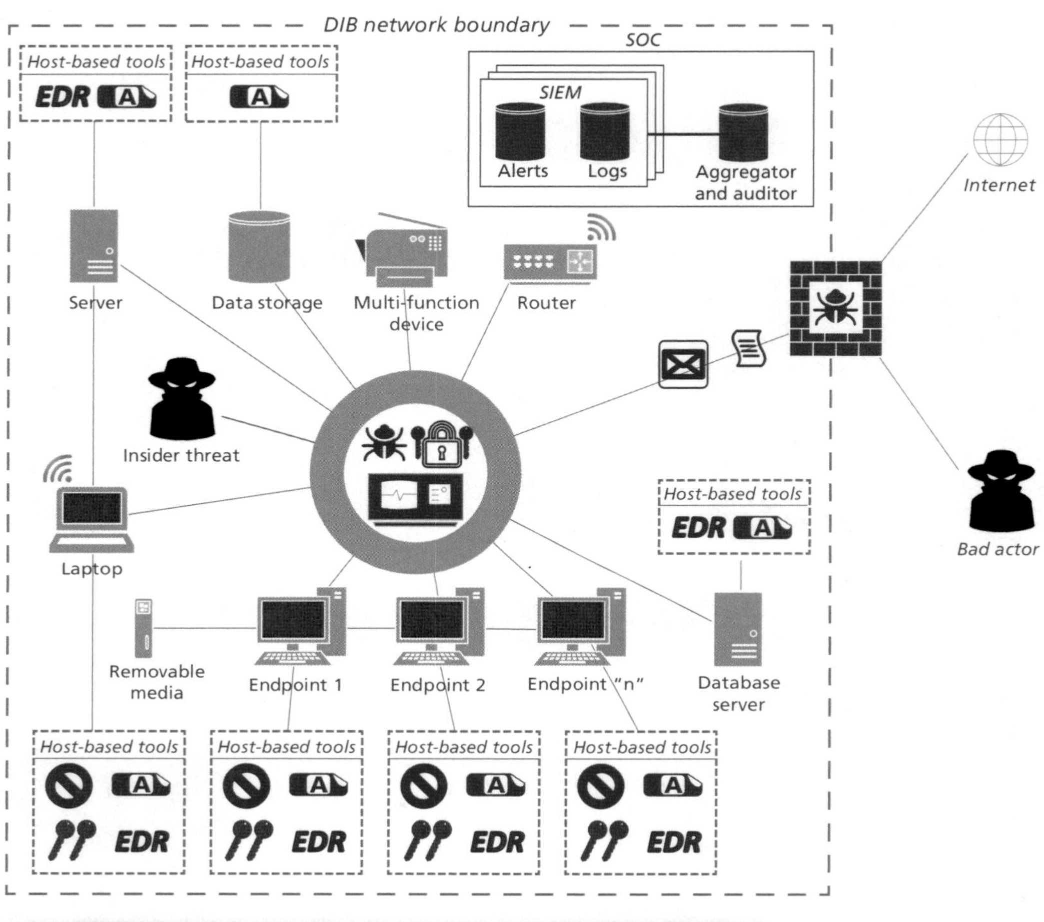

DIB firm CSTs					
	Advanced firewall		Endpoint antivirus/ antimalware	EDR	Endpoint detection and response tool
	Intrusion detection system (IDS)		Automated patching tool		Cyber threat intelligence
	Email security tools		Data-filtering app		Network access control (MFA)

NOTE: SOC = security operations center.

Figure A.2
Network Diagram for a Small DIB Firm (Status Quo)

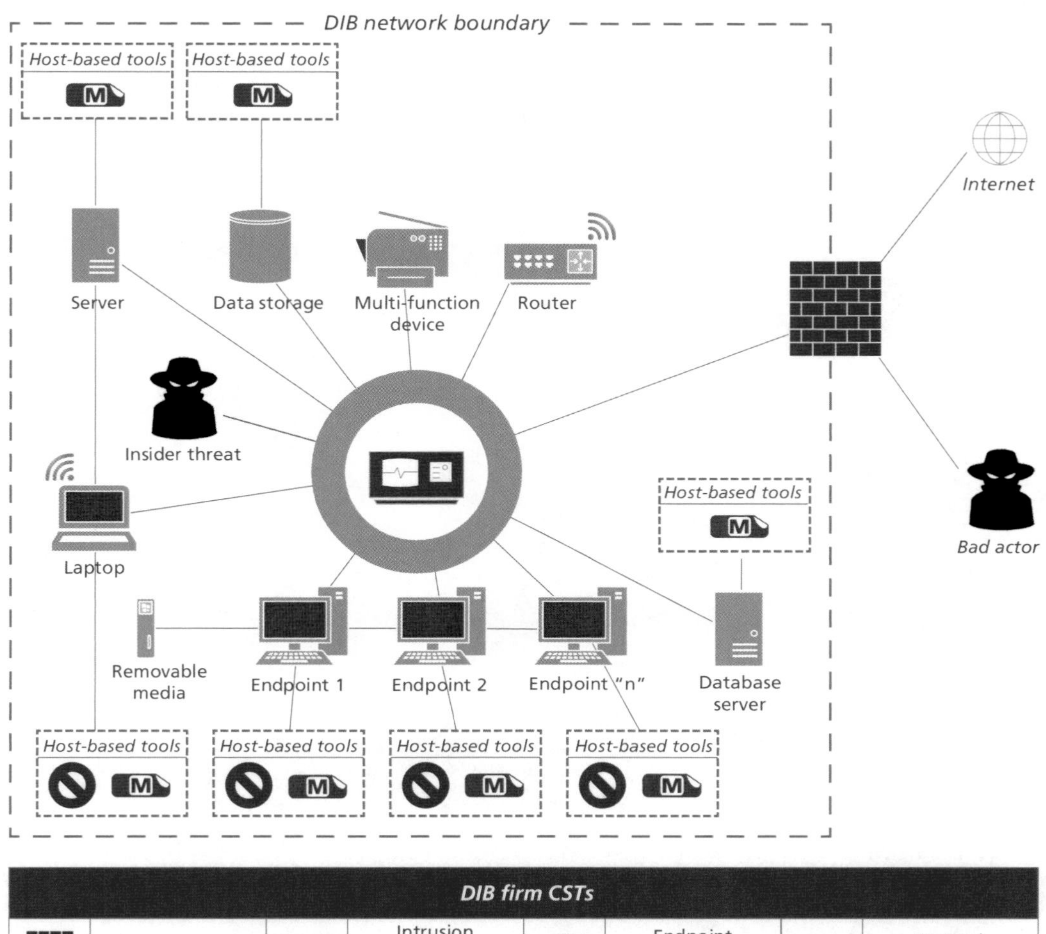

Option A: DoD-Lead Defense Industrial Base Security Operations Center

Figure A.3
Network Diagram for a Large Firm (Option A)

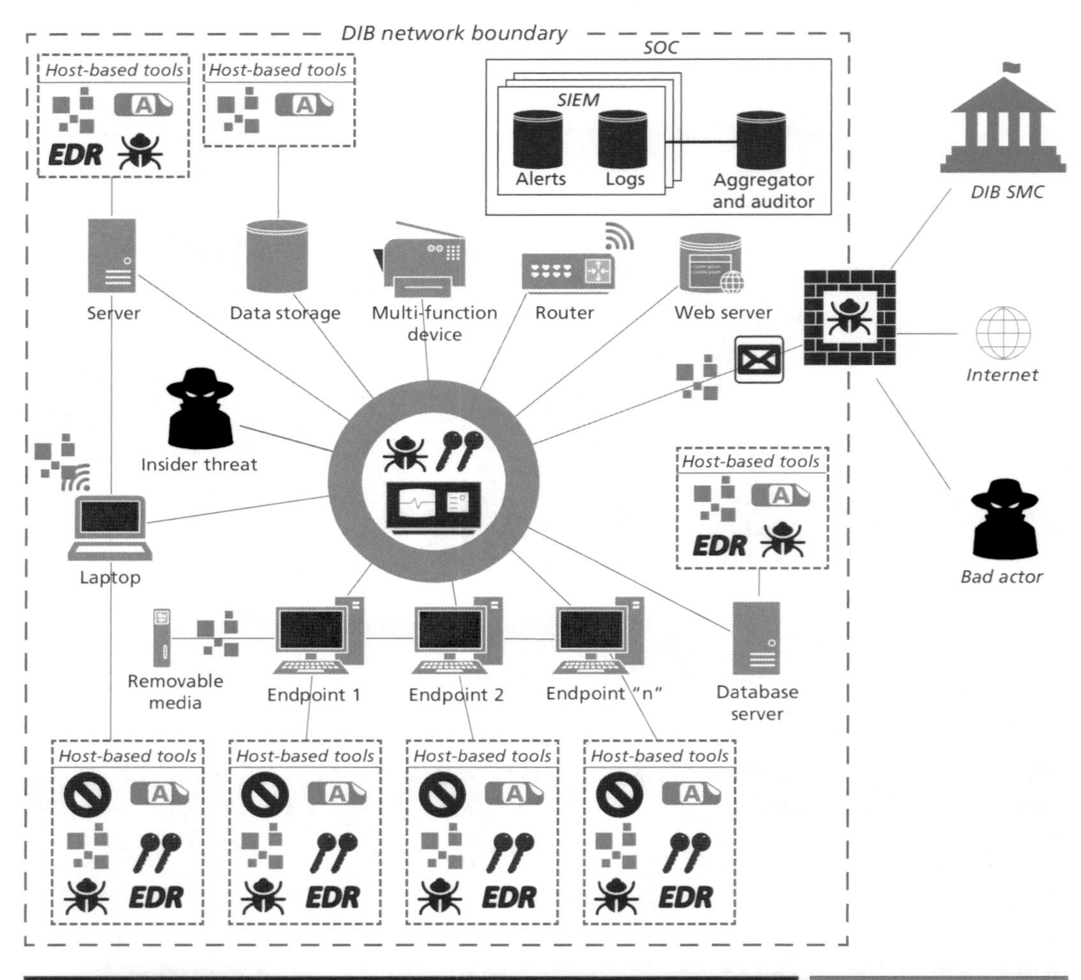

DIB firm CSTs						DCP2 CSTs	
	Advanced firewall		Endpoint antivirus/ antimalware		Cyber threat intelligence		Data loss prevention (DLP) app *MV CUI firms only*
	Intrusion detection system (IDS)	EDR	Endpoint detection and response tool		Network access control (MFA)		Automated patching tool
	Email security tools						Web security tools

Figure A.4
Network Diagram for a Large Firm with Cloud (Option A)

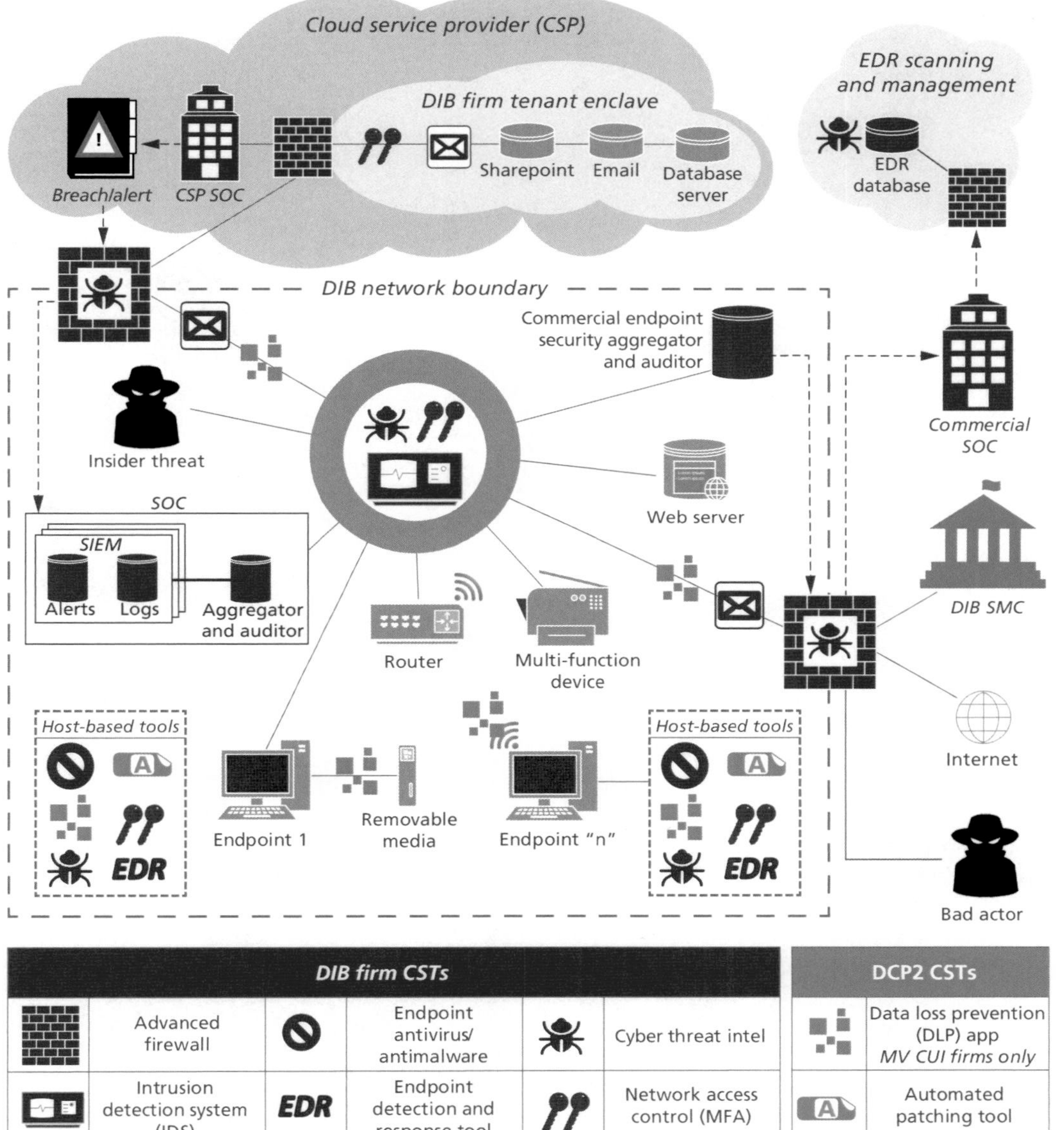

Figure A.5
On-Premises Network Diagram for a Small DIB Firm with High-Value Controlled Unclassified Information (Option A)

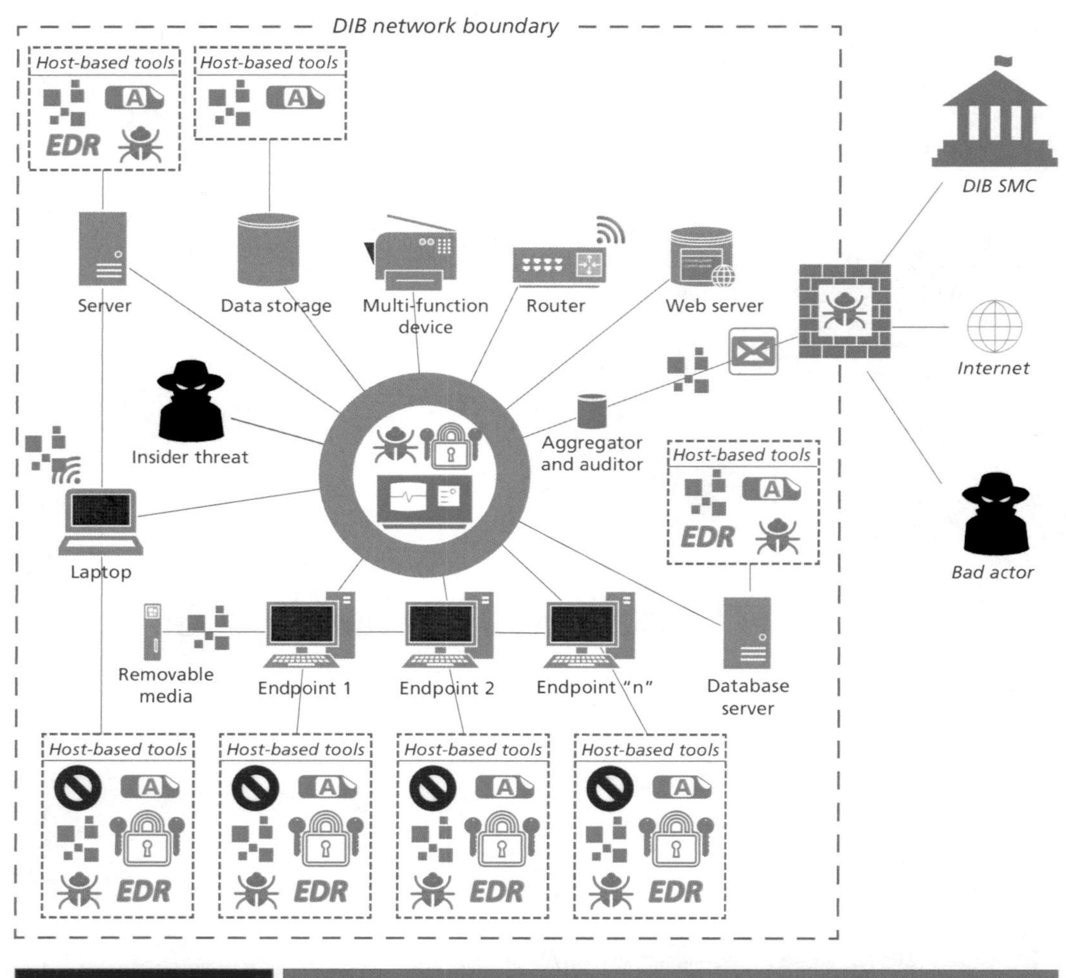

DIB firm CSTs	
	Endpoint antivirus/ antimalware

DCP2 CSTs					
	Advanced firewall		Data loss prevention (DLP) app *MV CUI firms only*	EDR	Endpoint detection and response tool
	Intrusion detection system (IDS)	A	Automated patching tool		Cyber threat intelligence
	Email security tools		Cryptographic secure MFA *MV CUI firms only*		Web security tools

Figure A.6
On-Premises Network Diagram for a Small DIB Firm with Moderate-Value Controlled Unclassified Information (Option A)

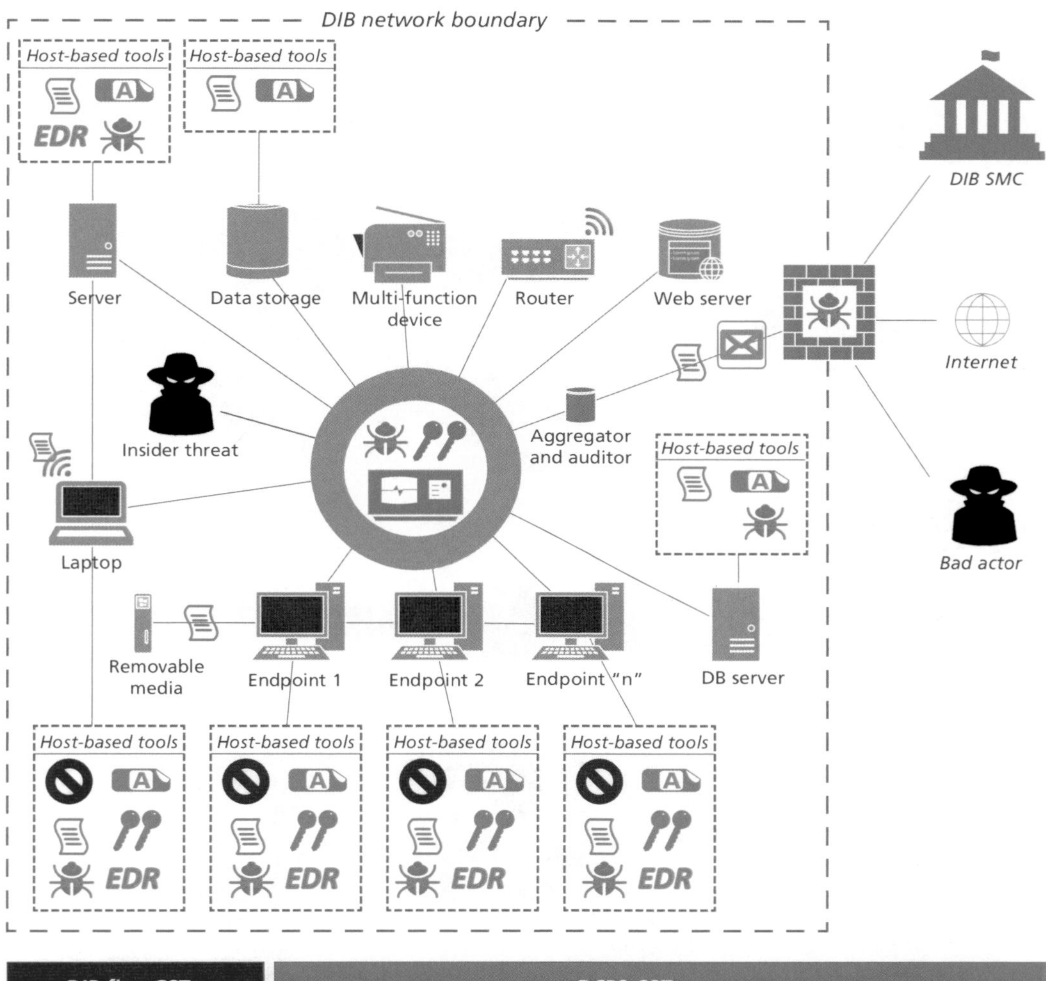

DIB firm CSTs	
	Endpoint antivirus/antimalware

DCP2 CSTs					
	Upgraded firewall		Data-filtering app *MV CUI firms only*	EDR	Endpoint detection and response tool
	Upgraded intrusion detection system (IDS)		Automated patching tool		Cyber threat intelligence
	Email security tools		Network Access Control (MFA) *MV CUI firms only*		Web security tools

Figure A.7
Cloud Network Diagram for a Small DIB Firm with High-Value Controlled Unclassified Information (Option A)

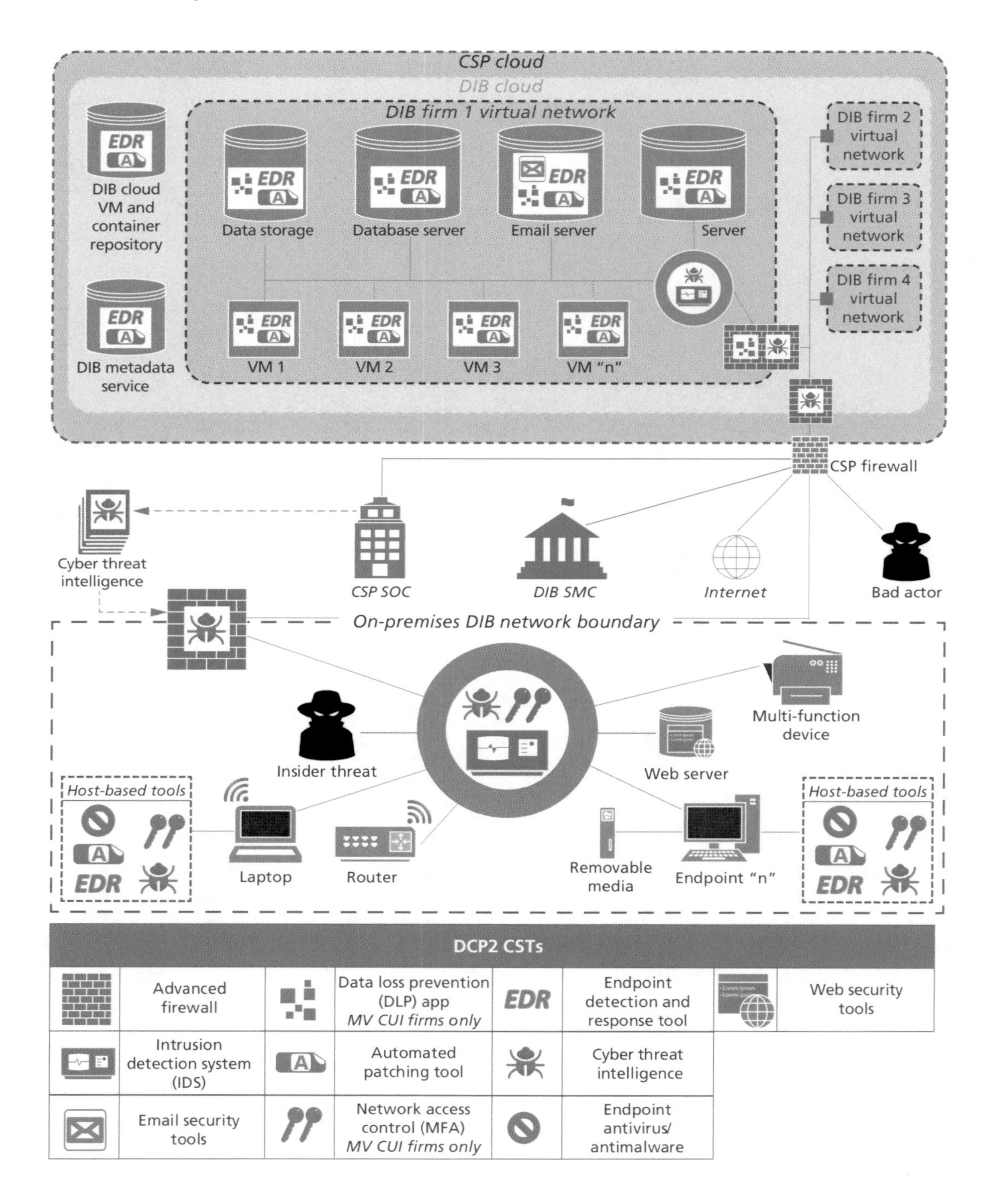

Figure A.8
Cloud Network Diagram for a Small DIB Firm with Moderate-Value Controlled Unclassified Information (Option A)

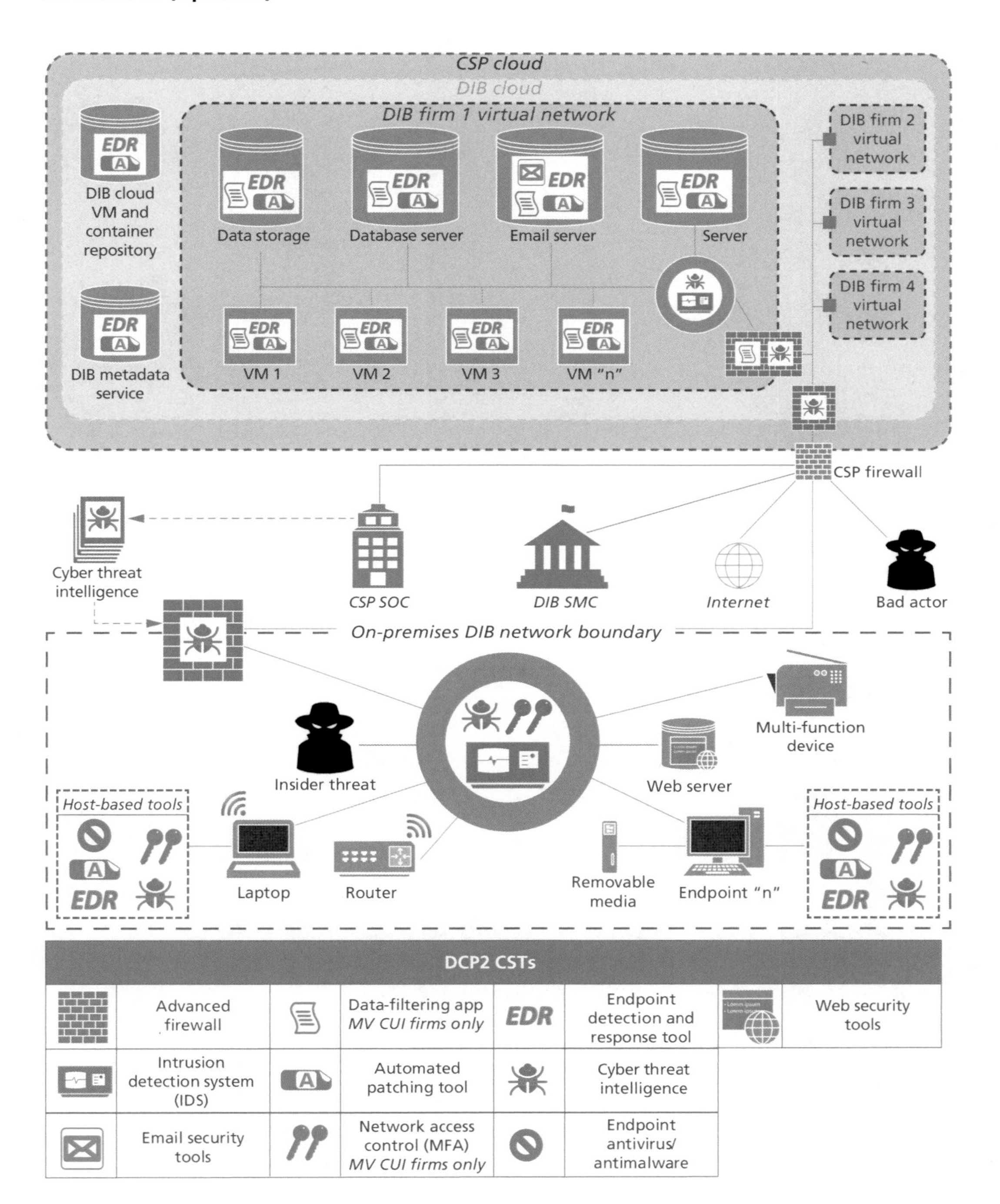

Option B: Commercial-Lead Security Operations Center

Figure A.9
Network Diagram for a Large Firm (Option B)

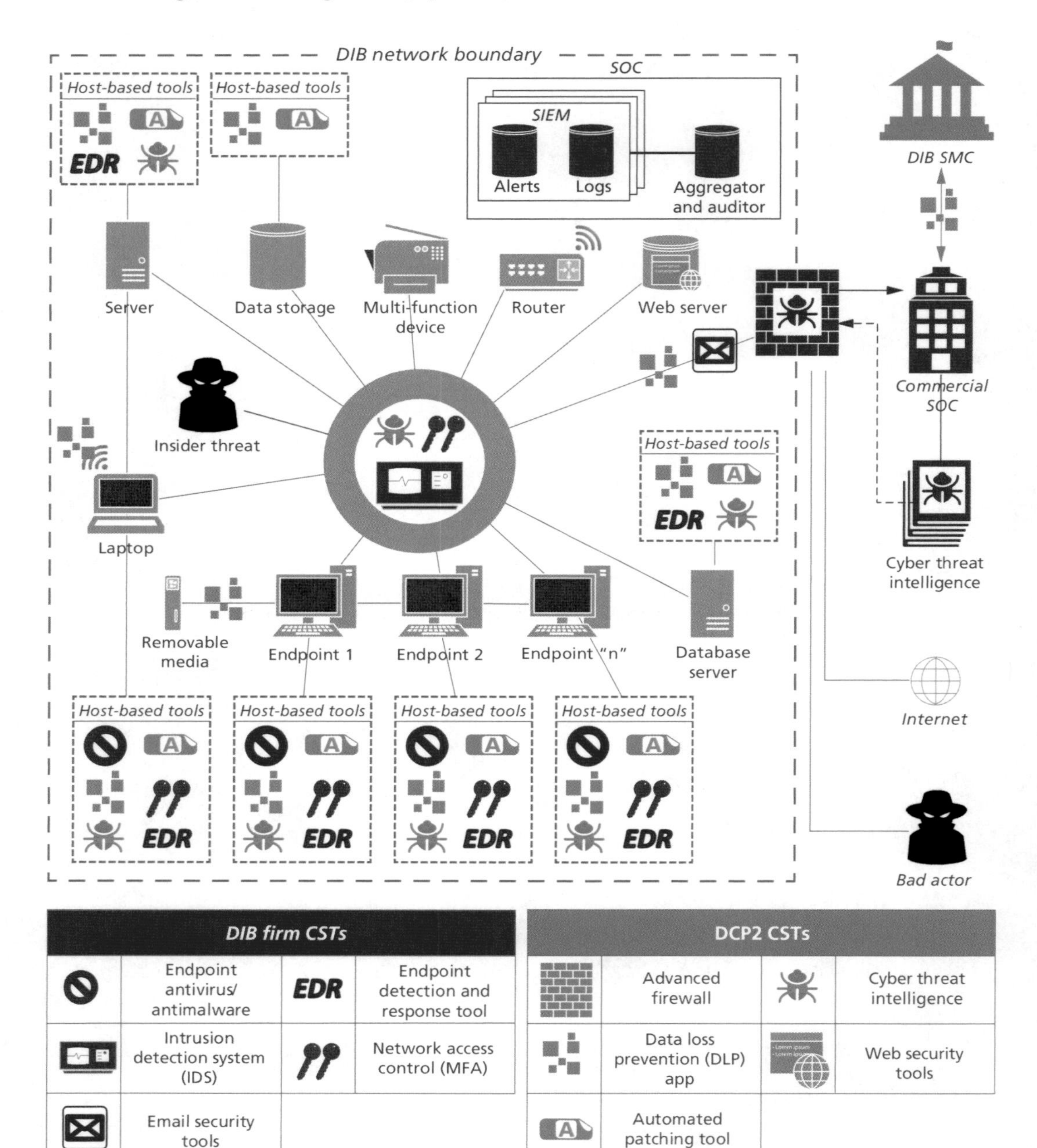

Figure A.10
On-Premises Network Diagram for a Small DIB Firm with High-Value Controlled Unclassified Information (Option B)

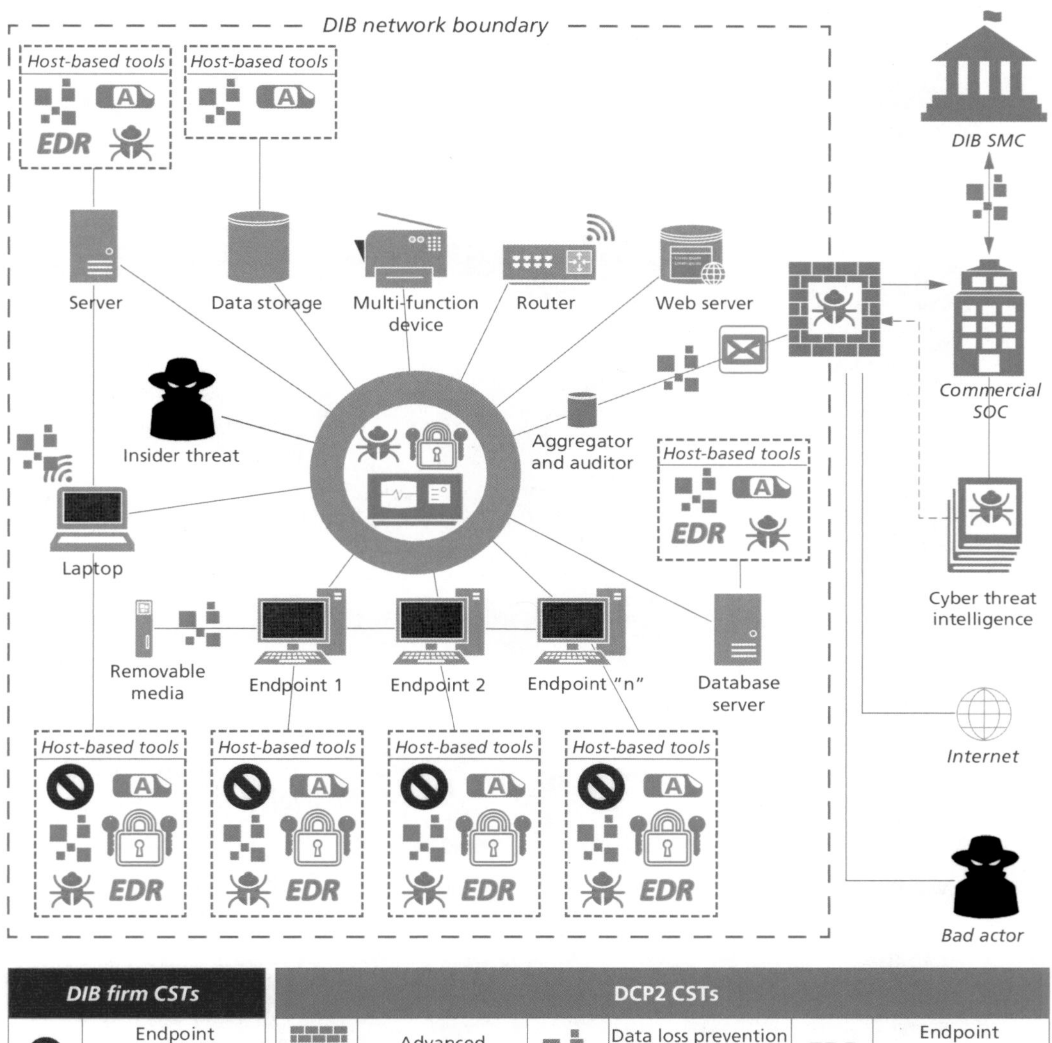

DIB firm CSTs	
	Endpoint antivirus/ antimalware

DCP2 CSTs					
	Advanced firewall		Data loss prevention (DLP) app *MV CUI firms only*	EDR	Endpoint detection and response tool
	Intrusion detection system (IDS)		Automated patching tool		Cyber threat intelligence
	Email security tools		Cryptographic secure MFA *MV CUI firms only*		Web security tools

Figure A.11
On-Premises Network Diagram for a Small DIB Firm with Moderate-Value Controlled Unclassified Information (Option B)

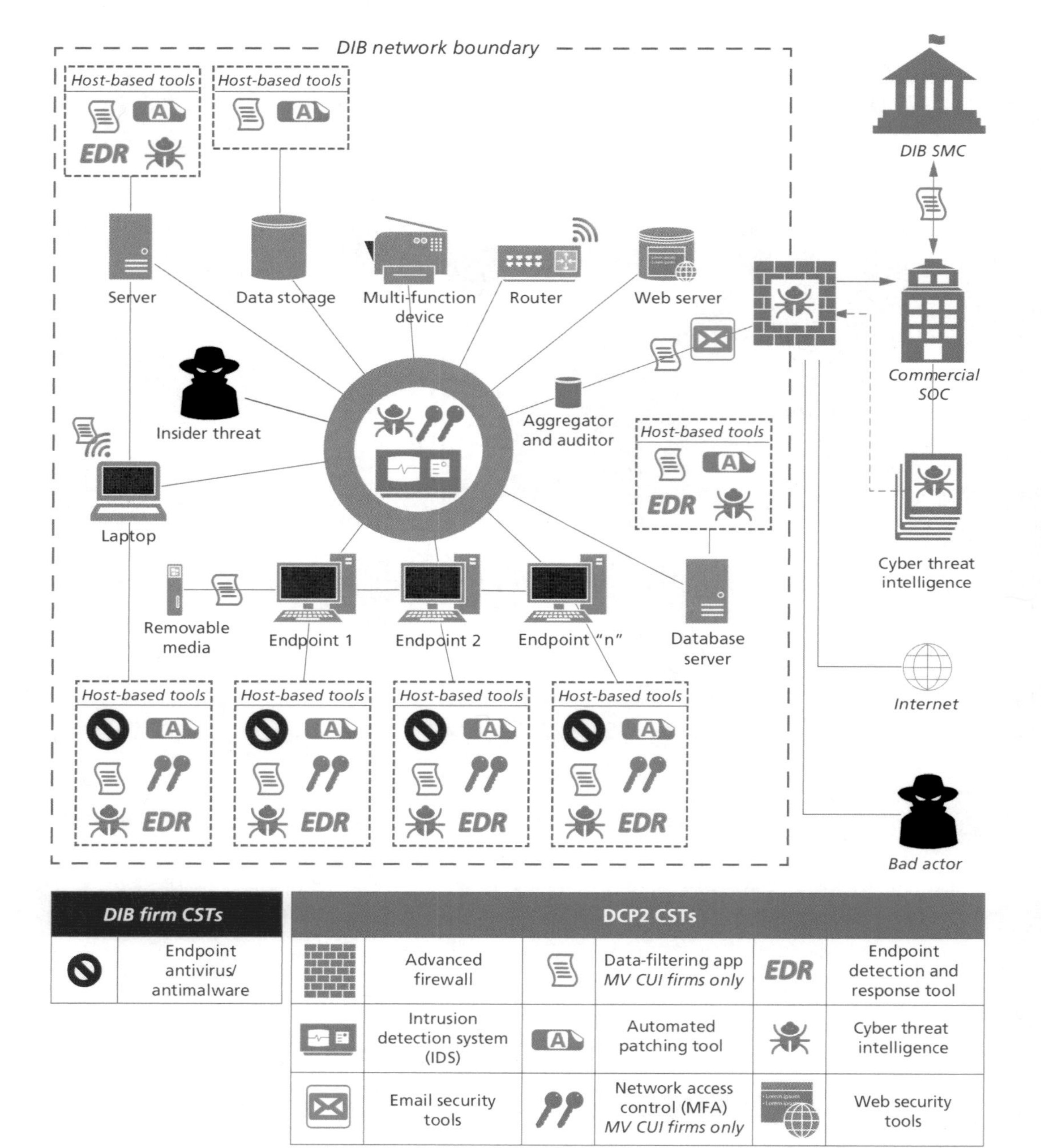

Figure A.12
Cloud Network Diagram for a Small DIB Firm with High-Value Controlled Unclassified Information (Option B)

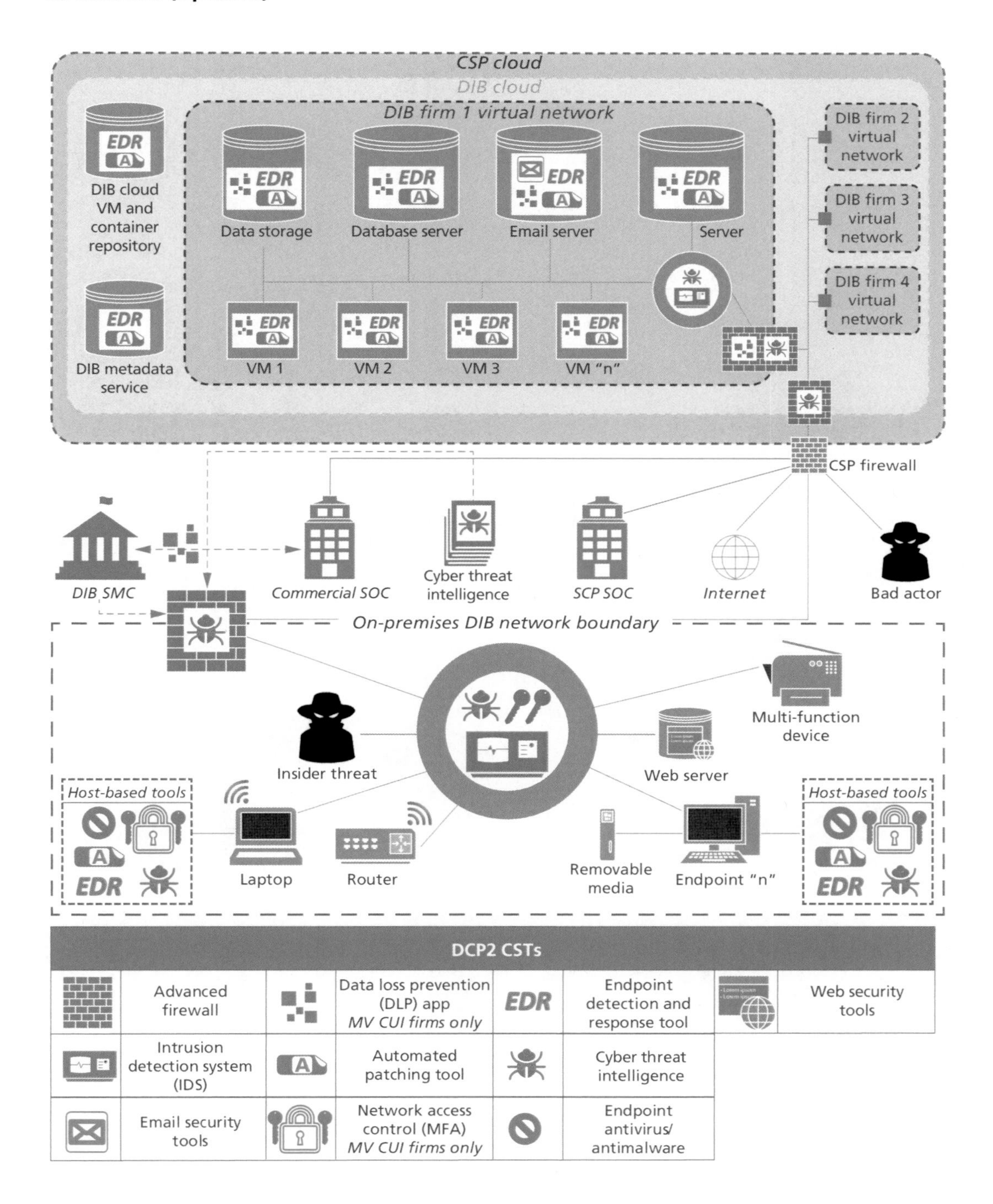

Figure A.13
Cloud Network Diagram for a Small DIB Firm with Moderate-Value Controlled Unclassified Information (Option B)

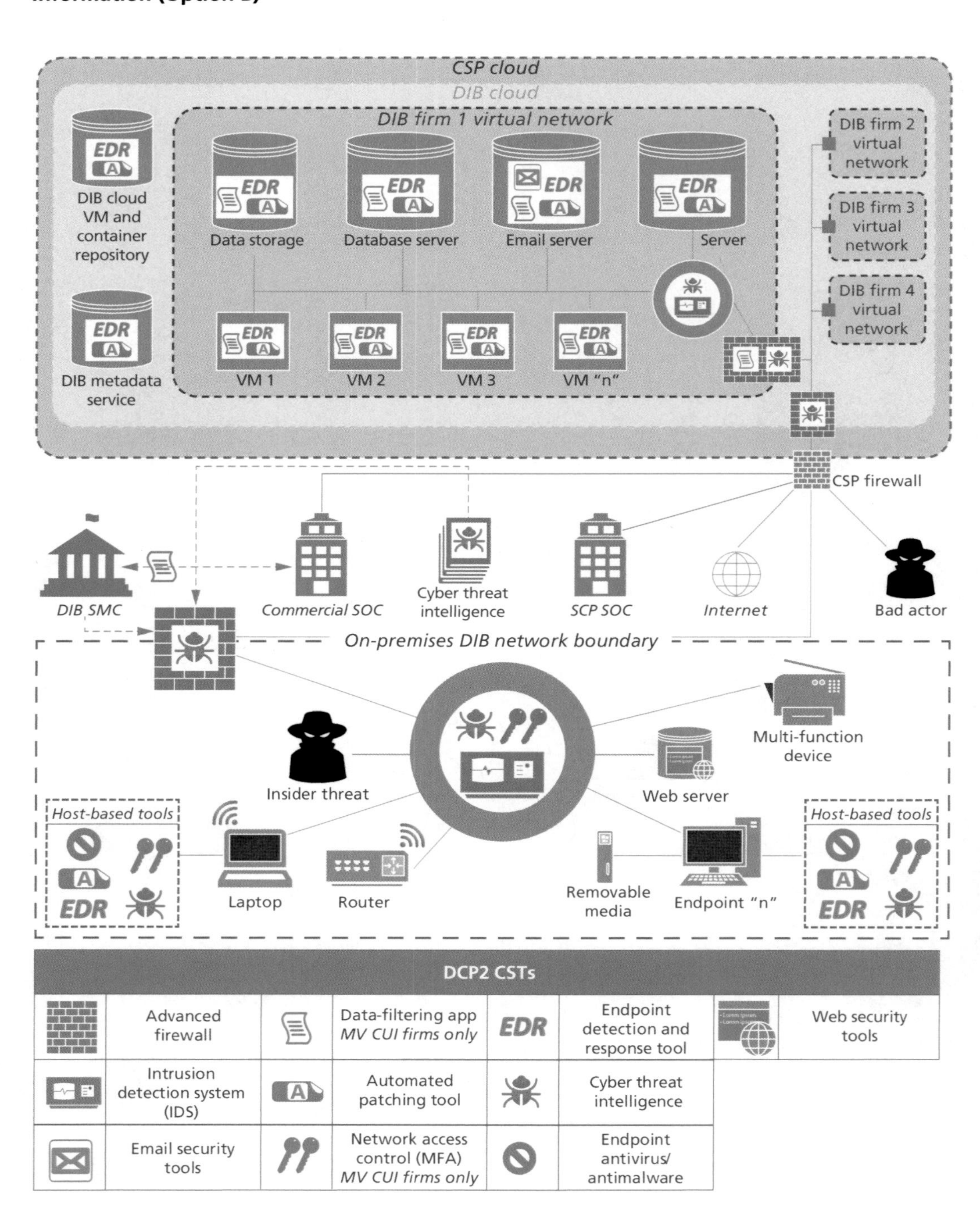

APPENDIX B

Cybersecurity Tools from Select Cybersecurity Firms

In this appendix, we summarize the types of CSTs available today in the market and then describe in detail the capabilities of CSTs offered a few vendors. We held discussions with several commercial cybersecurity firms to better understand how their tools and their capabilities could be used to improve the cyber defense of DIB firms. In this appendix, we also categorize CSTs so we can map them to the proposed DCP2.

We used a number of sources to identify CST tool categories and vendors that offer industry leading CSTs. For example, we used the 2019 Gartner Magic Quadrant report for Managed Security Services to identify leading vendors in the managed security services category.[1] We used other Gartner reports and other industry sources to identify other tool categories and industry leaders in these categories.

Table B.1 lists the tool categories used by industry analysts and CST vendors that offer industry leading tools in each category. CST categories change over time as markets and technologies change. There is also some overlap in the capabilities offered by CSTs in different categories. For example, there is overlap in the capabilities provided by antivirus endpoint security systems and EDRs, between EDRs and DLP applications, and between SIEM and managed security services. For the most part, the CST categories used in this report follow standard industry practice, except for the following exceptions. In this report, we combine two categories: (1) managed security services and (2) SIEM; we call this combined system capability the system component of a security operations center (SOC).

It should be noted that the network access control category is a broad category that incorporates many different technical approaches and technologies. It also is referred in different ways by cybersecurity vendors and industry analysts: *user access control, network access control,* and *user authentication*. The cybersecurity industry typically bundles all of these together into a single category, and we do as well, and call it

[1] Gartner, *Magic Quadrant for Managed Security Services, Worldwide (ID: G00354867)*, May 2, 2019. Gartner is a widely respected market research IT consulting firm that analyzes market trends, vendor product maturity, and market participants to identify worldwide industry leaders in information technologies and cybersecurity services.

network access control (for this reason, the icons for the two capabilities in Table B.1 are the same).

Table B.1 summarizes the tools and capabilities offered by specific vendors in each of these categories. These are the same tool categories used in the DCP2 cybersecurity architectures described in Chapter Six and in Appendix A, with one exception. The last tool category indicated in the table is not included in Appendix A. It is an emerging capability called *security validation*. Security validation tools provide an autonomous or semi-autonomous capability that can monitor and assess the overall cybersecurity status of a network and identify systems that require patching or may not be monitored adequately by deployed cybersecurity applications such as EDRs. This mapping is based on publicly available data and conversations held with representatives from several cybersecurity firms.

This appendix describes vendor products and capabilities in many of these CST categories. We contacted IBM, Secureworks, Symantec, Trustwave, and Verizon to discuss the managed security services and SIEM category. None of these responded, but we were able to learn more about CSTs in this category from FireEye, ArcSight, and Splunk. We also selected a number of other cybersecurity vendors that provide endpoint security tools and who are market leaders in this category. These firms also provide support to some DIB and government entities. Within this category of firms, we contacted Carbon Black, CrowdStrike, Cylance, FireEye, Palo Alto Networks, SentinelOne, and Trend Micro. Of the 12 total firms contacted, three responded: CrowdStrike, Cylance, and FireEye.

We discuss these tools and relevant companies in the subsequent sections; however, we will discuss DLP tools in appendix C. There are some overlaps in these tool categories, and not all vendor tools neatly fit into these categories. In the following sections, we discuss the CSTs offered by CrowdStrike, then those offered by Cylance, and finally those offered by FireEye. All these firms also provide managed security services in varying degrees—from general guidance on attack type, to complete monitoring, detection, response, and recovery. Although we were not able to meet with the market leaders in managed security services, we were able to learn about these from Cylance, CrowdStrike, and FireEye. We discuss their security service offerings below.

Some have speculated that the cybersecurity industry will experience significant consolidation in the future. More cybersecurity firms are offering a wide array of tools and services that span all of the categories in Table B.1.

CrowdStrike

Considered a market leader by both Gartner and Forrester, CrowdStrike has focused on cyber threats proliferated by nation-states and—aggregating these data—has generated a proprietary database of related threat intelligence particularly relevant to DIB

Table B.1
Cybersecurity Tool Categories and Vendors

Icon	Tool Category	Selected Relevant Companies
	Antivirus, endpoint security	Symantec, McAffee, CrowdStrike, Trend Micro, Carbon Black
	Advanced firewalls	Palo Alto Networks, Cisco, Fortinet, Check Point Software, Forcepoint
	Intrusion detection system (IDS)	Trend Micro, Cisco, McAfee, Forcepoint
	Data loss prevention (DLP)	Digital Guardian, Fidelis, Forcepoint
	Data-filtering applications	Palo Alto Networks, Cisco, Fortinet, Sophos
A	Automated patching tools	Solarwinds RMM, Microsoft SCCM, Tenable, Ninite Pro, PDQ Deploy
	Email security tools	Cisco, Fireeye, Trend Micro, Proofpoint, Forcepoint, Microsoft, Symantec
EDR	Endpoint detection and response (EDR)	CrowdStrike, Cylance, Fireeye, Carbon Black, Symantec, Fidelis
	Cyber threat intelligence	Alienvault, CrowdStrike, Fireeye, Cylance, Recorded Future, Trend Micro, Forcepoint
	Network access control (MFA)	Duo, Yubico, Onelogin, RSA, Gemalto
	Managed security services	Secureworks, Trustwave, IBM, Symantec, Verizon
	Security incident event management (SIEM)	Splunk, IBM, LogRhythm, Dell (RSA), Exabeam, McAfee, Securonix
	Web security tools	Symantec, Zscaler, Cisco, McAfee, Forcepoint, Trend Micro
	Security validation	Verdoin, AttackIQ, Cymulate, SafeBreach

firms. In addition to perimeter defense, CrowdStrike has focused on other capabilities of importance to DIB firms, such as post-breach and retroactive threat detection and response, as well as endpoint and end user monitoring for anomalous behavior. Once detected, data are sent to CrowdStrike's proprietary cloud platform for analysis, enabling the on-premises CST infrastructure to be minimized; they also have the ability to both accept and transmit data too external cyber and network security services and applications. CrowdStrike's cloud infrastructure is FedRAMP certified and uses U.S.-citizen-only cloud management.

CrowdStrike CSTs form an integrated suite of capabilities that can operate across mobile devices, laptops, desktop servers, containers, VMs, APIs, applications, and clouds—and across multiple operating systems, all of which encompasses the major components of DIB firms. The following is a summary of cybersecurity capabilities offered by CrowdStrike that are relevant to the DCP2.

Table B.2
Falcon Prevent

	Description and Capabilities
Description	"Falcon Prevent" falls under the industry category of a next-generation antivirus.
Features	• Monitors for traditional malware and file-less intrusion events when endpoints are both connect and not connected to the internet. • Monitors for intrusions from commodity malware, zero-day malware, and malware-free attacks. • Utilizes machine learning, exploit mitigation, whitelisting/blacklisting, and indicators of attacks to identify and mitigate breaches.

SOURCE: CrowdStrike, "Next-Generation Antivirus (NGAV): Falcon Prevent," 2019f.

Table B.3
Falcon Insight

	Description and Capabilities
Description	"Falcon Insight" monitors endpoints connected to the network and records event activity for inspection.
Features	• Event activity that is recorded can be viewed in real time or reviewed for future investigations. • Utilizes a proprietary cloud-based analysis engine that leverages AI and machine learning on endpoint event data received to monitor and identify attacks. • Prioritizes alerts for security teams. • Can provide information on attacks attributed to specific adversaries. • Provides controls to contain compromised systems.

SOURCE: CrowdStrike, "Endpoint Detection & Response: Falcon Insight," 2019b.

Table B.4
Falcon Device Control

	Description and Capabilities
Description	"Falcon Device Control" manages the interaction of removable media with the network to control which devices can make and accept data transfers.
Features	• Does not require software installation on endpoints. • Allows testing for policy impact prior to enforcement. • Provides a list of devices along with identifying properties (e.g., vendor, product, serial number). • Can control device access based on device type, level of access, functions, user, etc. • Monitors and logs removable media usage history.

SOURCE: CrowdStrike, "Endpoint USB Device Control: Falcon Device Control," 2019c.

Table B.5
Falcon OverWatch

	Description and Capabilities
Description	"Falcon Overwatch" is a threat hunting tool that identifies threats in the network and pushes the threat information (e.g., tactics, techniques, and procedures) up to the global CrowdStrike community.
Features	• Actively monitors for threats to the network. • Ranks threat alerts for response teams to review. • Can be linked to an external security team for guidance on attack response.

SOURCE: CrowdStrike, "Proactive Threat Hunting: Falcon OverWatch," 2019g.

Table B.6
Falcon Discover

	Description and Capabilities
Description	"Falcon Discover" scans the network to identify attached assets, existing credentials, and endpoint configurations.
Features	• Identifies which applications are running on which hosts and captures when each application was originally launched. • Identifies usage by application or by host. • Gathers list of administrator credentials by user and corresponding usage history help identify inappropriate use of access. • Generates real-time application inventory along with usage to highlight potential cost-savings opportunities.

SOURCE: CrowdStrike, "Network Security Monitoring & IT Hygiene: Falcon Discover," 2019e.

Table B.7
Falcon Spotlight

	Description and Capabilities
Description	"Falcon Spotlight" monitors and assesses the vulnerability of endpoints associated with the network.
Features	• Monitors endpoints both on and off the network. • Combines and correlates events for review by security operators. • Does not require on-premises infrastructure.

SOURCE: CrowdStrike, "Fast, Effective Vulnerability Assessment: Falcon Spotlight," 2019d.

Table B.8
Falcon X

	Description and Capabilities
Description	"Falcon X" integrates cyber threat intelligence and endpoint protection.
Features	• Automates incident investigations and breach responses through pre-approved controls. • Learns from past attacks to better tailor future mitigation strategies. • Based on the situation, quarantines files automatically. • Performs malware file analysis in a cloud-based sandbox. • Groups and correlates malware events to highlight malware families and related campaigns. • Suggests tactics, techniques, and procedure attribution and countermeasures based on cyber threat intelligence. • Cloud delivered, does not require on-premises infrastructure.

SOURCE: CrowdStrike, "Cyber Threat Intelligence Platform: Falcon X," 2019a.

Cylance

Cylance—another FedRAMP-certified CST vendor—has focused on predicting and preventing attacks by leveraging AI and developing proprietary machine learning algorithms. Cylance takes a known malware sample and creates multiple hashes of different sections of the same piece of malware; if this malware is reintroduced to the network—whether in its entirety or only a recycled portion—the known hash will trigger a response from the CSTs.

While Cylance's capabilities are designed to integrate with one another, they can also cohabitate with other, preexisting security controls. The following is a summary of cybersecurity capabilities offered by Cylance which are relevant to the DCP2.

Table B.9
CylanceGUARD

	Description and Capabilities
Description	"CylanceGUARD" is an EDR module that works at the network level.
Features	• Uses AI to detect known and zero-day threats, polymorphic malware, APTs, and both file-based and file-less threats. • Uses intelligence and methodology-based processes to identify potential attacks based on previous patterns, data exfiltration, and unauthorized access. • Logs event details and provides suggested countermeasures. • Collates event data and ranks alerts by severity. • On-boarding is supported by a Cylance response team.

SOURCE: BlackBerry Cylance, "CylanceGUARD," 2019a.

Table B.10
CylancePROTECT

	Description and Capabilities
Description	"CylancePROTECT" monitors the endpoints for attacks and can automate response mechanisms.
Features	• Utilizes AI to detect attacks before a malicious script can execute. • Can lockdown specified systems when triggered by a breach and prevent changes thereafter until assessed by a security operator. • Sets parameters for what devices can connect to the environment; access can be very specific (e.g., specific serial numbers) or generic (e.g., device category).

SOURCE: BlackBerry Cylance, "CylancePROTECT," 2019c.

Table B.11
CylanceOPTICS

	Description and Capabilities
Description	"CylanceOPTICS" is an endpoint detection and response solution that uses AI to detect widespread security incidents.
Features	• Detection and response decisions occur at the endpoint, reducing response latency. • Rules can be configured to a specific playbook, initiating a discrete set of responses given the type of event trigger. • Includes threat hunting tools. • Includes tools to analyze prevented attacks. • Event logs are archived in a standard format. • Can execute a partial lockdown, allowing security operators to maintain communication with a suspected compromised endpoint.

SOURCE: BlackBerry Cylance, "CylanceOPTICS," 2019b.

Table B.12
Cylance Smart Antivirus

	Description and Capabilities
Description	"Cylance Smart Antivirus" falls under the industry category of a next-generation antivirus.
Features	• Compatible with various operating systems. • Uses AI and predictive analytics to mitigate breaches.

SOURCE: BlackBerry Cylance, "World-Class Antivirus Protection," 2019d.

FireEye

Considered a market leader by both Gartner and Forrester, FireEye has been collecting threat intelligence for many years, generating cyber intrusion sets on a large number of nation-state sponsored APTs—relevant attackers of DIB firms. Leveraging its global network, FireEye claims to collect more than 600,000 malware samples every day, which enables it to generate behavior-based solutions rather than solely signature-based ones.

The FireEye ecosystem is an integrated suite of detection, protection, and investigation capabilities. The following is a summary of cybersecurity capabilities offered by FireEye that are relevant to the DCP2.

Table B.13
Helix Security Platform

	Description and Capabilities
Description	FireEye's "Helix Security Platform" integrates several FireEye security tools and augments them with SIEM, orchestration, and threat intelligence capabilities to mitigate, detect, and respond to security events.
Features	• Identify existing breaches. • Automates alert validation, given pre-approved controls. • Provides an overlay of cyber threat intelligence and threat analytics for security teams. • Provides compliance reporting dashboards. • Can automate workflows based on predesign playbooks.

SOURCE: FireEye, "Helix Security Platform," 2019.

Table B.14
Endpoint Security

	Description and Capabilities
Description	FireEye's "Endpoint Security" provides endpoint detection and response for attacks.
Features	• Uses multi-level detection, combining signature-based, behavior-based, and intelligence-based indicators of compromise. • Retains endpoint activity for forensic investigations. • Can conduct searches of all network endpoints to locate compromised devices, and deploy the same response to each identified endpoint. • Uses a proprietary machine learning algorithm to detect malware.

SOURCE: FireEye, "Endpoint Security," 2019.

Table B.15
Network Security and Forensics

	Description and Capabilities
Description	FireEye's "Network Security and Forensics" monitors for attacks at the network level.
Features	• Can identify multi-flow, multi-stage, zero-day, polymorphic, and ransomware attacks. • Performs retroactive attack detection. • Performs real-time attack detection. • Can automate event triage through signature-based detection. • Uses a proprietary algorithm to prioritize response resources.

SOURCE: FireEye, "Network Security and Forensics," 2019.

Table B.16
SmartVision

	Description and Capabilities
Description	FireEye's "SmartVision" is a network traffic analysis (NTA) tool solution that detects suspicious lateral-moving traffic within a network.
Features	• Combines an analytics engine, machine learning module, and intrusion detection rules to identify indicators of compromise and data exfiltration attempts. • Scalable.

SOURCE: FireEye, "Network Security and Forensics," 2019.

Table B.17
Email Security

	Description and Capabilities
Description	FireEye's "Email Security" blocks malware, phishing URLs, and impersonation techniques.
Features	• Automatically blocks unknown malicious attacks. • Inspects URLs for links to credential-phishing sites and rewrites URLs. • Detects malware-less attacks. • Detects and alerts on URLs that "go live" after email delivery. • Remove emails from a user's inbox that become malicious after delivery. • Integrates with cloud-based email systems.

SOURCE: FireEye, "Email Security," 2019.

APPENDIX C

Data Loss Prevention Tools

DLP tools are "a set of tools and processes used to ensure that sensitive data is not lost, misused, or accessed by unauthorized users."[1] DLP capabilities can also help protect data from being erased or destroyed by malicious actors that use malware, such as ransomware. In addition to *data loss prevention*, these capabilities are often referred to by several other terms: [2]

- data leakage prevention
- extrusion prevention (EP)
- content monitoring and filtering (CMF)
- content monitoring and protection (CMP)
- information leakage prevention (ILP)
- outbound content compliance (OCC)
- information protection and control (IPC).

An effective DLP solution must address the state-specific vulnerabilities data may encounter as it is generated, transformed, and stored. These states include:[3]

- **Data in motion (DIM):** Also referred to as "data in transit"; data actively moving from one location to another.
- **Data in use (DIU):** Data that are actively being used and transformed.
- **Data at rest (DAR):** Data that are not actively being moved or used (i.e., data stored on a hard drive for future use or archiving purposes).

---

[1] Digital Guardian, *Endpoint Data Loss Prevention*, 2019.

[2] Barbara Hauer, *Data Leakage Prevention: A Position to State-of-the-Art Capabilities and Remaining Risk*, Institute of Systems Software, 2014.

[3] Nate Lord, "Data Protection: Data In Transit vs. Data at Rest," *Digital Guardian*, January 3, 2019; Hauer, 2014.

Aspects of Effective DLP Software

We found that, in order for a DLP solution to be effective—given the specific intent for its acquisition—it must address one or more of the seven aspects listed below. A truly comprehensive, robust DLP solution will address most, if not all, seven aspects.

Data Discovery

This aspect of a DLP solution is premised on the idea that you cannot protect data that you do not know exists and it provides a capability to identify the location and type of each bit of data on an entity's network. This is achieved in three steps. First, the DLP software scans the entire network, identifying both structured and unstructured data. Next, as data are located, they are fingerprinted and tagged as a particular form of data (e.g., a purchase contract), if possible; these "data forms" are set up in advance, and many data-discovery DLP solutions come with commonly used data forms already configured. Finally, once the data have been fingerprinted and tagged, each identified bit of data is categorized and classified in terms of sensitivity.

Endpoint Protection

This aspect of a DLP solution focuses on monitoring and enforcing rules on end-user actions. With this solution, sensitive data (e.g., IP, PII) are protected by identifying and monitoring potential leak channels. Common channels for data leakage include

- removable storage devices (e.g., USB, DVD)
- virtual desktops
- cloud applications (e.g., Dropbox, Google Drive)
- email applications (e.g., Outlook)
- filesharing applications
- instant messaging applications
- web browsers
- network protocols (HTTP, HTTPS, FTP)
- network file share
- social media
- printing and faxing (including images saved to a clipboard and/or screen captures).

One means by which endpoint protection DLP solutions work is by focusing on device control via a set of enterprise rules—some of which may come as defaults, and some of which can be set up after acquisition. These rules generally emphasis the control of data transfer at three distinct levels:

- **Employee level:** At this rule level, a DLP solution monitors and regulates which employees within an enterprise can access and transfer certain categories of data.

For example, a DLP solution may allow employees assigned to an accounting department to access payroll data but prevent all other employees from accessing that same data.

- **Endpoint-type level:** At this rule level, a DLP solution specifies the endpoint types within an enterprise that can and cannot be used for transferring each category of data. For example, a DLP solution may allow data categorized as "internal research reports // for public release" to be transferred to any type of removable device (e.g., USB) but prevent data categorized as "internal research reports // FOUO" from being transferred to those same removable devices.
- **Data-category level:** At this rule level, a DLP solution enforces enterprise rules that prevent certain categories of data from being transferred all together. For example, a DLP solution may prevent data categorized as "contracts" from being transferred in any capacity.

Network Monitoring

This aspect of a DLP solution focuses on specifically monitoring an enterprise's network for sensitive data and enforcing enterprise rules on any data identified. These solutions have the ability to prevent sensitive data from being transferred over a wide range of communication protocols, including

- email (both internal and external)
- web-based instant messaging
- social media
- encrypted traffic
- FTP
- IPv6
- generic TCP.

When a sensitive bit of data is identified by a network monitoring DLP solution as violating a preset enterprise rule, the DLP software can execute one of three solutions in real time to prevent unauthorized data transmissions (based on the rule being violated):

- **Modify data:** In certain cases, a DLP solution may be able to modify the bit of data being transferred—effectively removing the sensitive portion, while allowing the remainder of the data to be transferred. A common example of this is the automatic removal of PII—such as social security numbers—from both internal and external emails.
- **Block data:** In the case where a DLP solution identifies sensitive information that is not cleared for transfer over any means (such as source code for special programs), the software can be set up to automatically block the data's transfer.

- **Redirect data:** In other cases, a DLP solution may identify data that (1) violate a specific rule, and/or (2) may not completely satisfy another rule; given the rule which is being violated or not satisfied, a DLP solution can chose to redirect the data to a holding location. Often, this holding location is a quarantine folder (as is the case for incoming messages identified as potentially harboring malware) or a review folder where either a supervisor or IT professional may go to review the bit of data and then determine whether it should be cleared for transfer or be blocked completely.

Another key feature of many DLP solutions is the ability to actively educate end users. When an end user is about to violate an enterprise data transfer or data access rule, the DLP solution can notify said end user that the action they are trying to perform violates protocol, along with an explanation as to why. Once notified, certain DLP solutions will allow the end user to acknowledge that they are aware they are violating protocol and proceed with the action—which can then be used to develop cases against users who are potential insider threats.

Data Storage Monitoring

This aspect of a DLP solution focuses on scanning and monitoring storage locations connected to an enterprise's network. This includes scanning the entire storage infrastructure for risks to data-at-rest, which includes

- file servers
- distributed machines (e.g., laptops)
- document/email repositories
- web content and applications (e.g., intranet)
- databases.

If a risk is identified during a scan, the DLP solution executes a pre-approved remediation action. Depending on the risk, these actions include file quarantine, file relocation, and policy-based file encryption.

Points-of-Egress Monitoring

This aspect of a DLP solution protects an enterprise's network at the external touchpoints by ensuring enterprise-wide data security policies are enforced out to the network boundary; this includes the integration of malware detection into network firewalls. If a data protection policy violation is detected by the DLP software, the data transmission can be encrypted, redirected, and/or quarantined for later review by a supervisor or IT professional; in certain cases, such transmissions may be blocked entirely.

Cloud Monitoring

This aspect of a DLP solution monitors for sensitive data being transferred to and from the cloud by enforcing data policies on content sent to and extracted from cloud-based applications. Certain cloud monitoring DLP solutions also offer specialized capabilities to protect data residing in the cloud, such as the ability to "un-share" sensitive files sent to the cloud for sharing with both internal and external parties, as well as identity-based encryption that prevents data shared with a specific third party from being sent to other individuals and/or locations without your permission.

Insider Threat Monitoring

This aspect of a DLP solution monitors end-user use of enterprise data. Through the use of data policies assigned to both data categories and employee types, an insider threat monitoring DLP solution can prevent end users from accessing data and executing actions that are outside their business processes. Similar to the network monitoring feature, the insider threat monitoring DLP software can be set up to notify an end user when they are attempting to perform an action which violates an enterprise data policy. The software can then require the end user to acknowledge that they are aware they are violating protocol before proceeding with the action. This information can then be saved and used to identify employees who may exhibit patterns of data policy violations—whether intentional or not.

Key Capabilities of DLP Firms

Throughout the course of this research, we contacted several private-sector firms that offer DLP solutions to evaluate how the DLP aspects identified throughout this research mapped to real DLP solutions being offered in the marketplace. To select DLP firms and products for further investigation, the research team utilized a report from Gartner, a respected IT consulting firm. Gartner's 2017 *Magic Quadrant for Enterprise Data Loss Prevention* report identified seven leaders in the DLP market, and we contacted each of them.[4] Of the seven firms contacted, only two responded—Fidelis Cybersecurity and Forcepoint.

Fidelis Cybersecurity

Fidelis Cybersecurity has designed an integrated DLP suite it calls the "Elevate Platform," which consists of three component systems: Fidelis Network, Fidelis Endpoint, and Fidelis Deception. These capabilities can work together or separately to provide

---

[4] Brian Reed and Deborah Kish, *Magic Quadrant for Enterprise Data Loss Prevention*, Gartner, February 16, 2017. Unfortunately, 2017 was the last year Gartner published this report. The firms contacted were CA Technologies, Fidelis Cybersecurity, McAfee, RSA, Symantec, Verdasys (which was acquired by Digital Guardian), and Websense (which was acquired by Forcepoint).

cyber security capabilities which use structured metadata to detect, hunt, and respond to inbound and insider threats, as well as mitigate data theft.[5] This is done by compiling hundreds of data attributes, known vulnerabilities, endpoint processes, and event data collected by the platform sensors across on-premises and cloud-based networks.[6] Tables C.1–C.3 summarize the "Elevate Platform" capabilities that are relevant to the DCP2.

Table C.1
Fidelis Network

	Description and Capabilities
Description	Fidelis Network monitors and analyzes network traffic to mitigate data loss. The same technology used to detect data theft is also used to detect and decode obfuscated malware hidden in plain sight.
Features	• Performs deep session inspection of all network traffic, including email and web-based traffic. • Capable of monitoring all port and protocol activity. • Capable of capturing various levels of metadata granularity, including network transaction details, hashes, and transferred-file properties. • Capable of detecting malware.

SOURCE: Fidelis Cybersecurity, 2019.

Table C.2
Fidelis Endpoint

	Description and Capabilities
Description	Fidelis Endpoint analyzes and records processes and events across endpoints.
Features	• Provides endpoint detection and response (EDR). • Provides forensic investigation tools to analyze activities. • Provides response and system management capabilities.

SOURCE: Fidelis Cybersecurity, 2019.

5 Fidelis Cybersecurity, "Executive Summary," undated.

6 Fidelis Cybersecurity, "Fidelis Elevate™," data sheet, April 2019.

Table C.3
Fidelis Deception

	Description and Capabilities
Description	Fidelis Deception uses information from the host network to generate decoy assets and embeds these decoys within the real network to lure attackers and malicious insiders, triggering a breach alert.
Features	• Automatically scans the host network to discover and classify network assets. • Identifies the services running on each network asset, and their level of connectivity. • Detects attacks and lateral movements within the network. • Captures network traffic and telemetry data for investigations. • Can be configured to automatically adjust decoys to match changes made to the real network. • Enables organizations planning to obtain CMMC Level 5 certification.

SOURCE: Fidelis Cybersecurity, 2019.

Forcepoint

Forcepoint has designed an integrated DLP suite composed of four modules: Endpoint, Cloud Applications, Discovery, and Network. Forcepoint's solution involves a user interface that can examine file transfer and copy events in real-time. Cybersecurity operators can apply a user-risk scoring scheme to these events to determine which events to approve and deny.[7] This system can also identify where sensitive data—such as critical business IPs and personal data—is on the network by scanning repositories and endpoints; once identified, the data are fingerprinted, and relevant endpoints are continuously monitored thereafter. Data-protection actions can be set up to automatically adjust as new information flows in (e.g., user and endpoint behavior) and also include a proprietary "drip" DLP dimension that implements a time-based algorithm that monitors for data loss over time (rather than in one large data exfiltration event). Tales C.4–C.7 summarize Forcepoint DLP capabilities that are relevant to the DCP2.

Table C.4
Forcepoint DLP Module: Endpoint

	Description and Capabilities
Description	Forcepoint's Endpoint DLP module monitors network endpoints to mitigate data loss.
Features	• Can be used on various operating systems, including MacOS, Windows, and Linux. • Capable of analyzing encrypted data and applying preselected DLP controls based on the results. • Includes a library of global regulations and corresponding controls for data compliance verification. • Incorporates an employee education component to inform employees of actions which violate enterprise data policies, and can request employee interaction to verify intent when interacting with sensitive data. • Monitors data transfers to the cloud and internet—including HTTPS.

SOURCE: Forcepoint, 2019.

---

[7] Forcepoint, *Forcepoint Data Loss Prevention (DLP): Data Protection in a Zero-Perimeter World*, 2019.

Table C.5
Forcepoint DLP Module: Cloud Applications

	Description and Capabilities
Basic Description	Forcepoint's "Cloud Applications" DLP module monitors cloud application activity to mitigate data loss.
Features	• Incorporates AI and analytics to monitors various cloud applications for data loss. • Ranks events based on a proprietary user-risk scoring scheme to highlight possible high-risk/high-loss events for security personnel.

SOURCE: Forcepoint, 2019.

Table C.6
Forcepoint DLP Module: Discovery

	Description and Capabilities
Basic Description	Forcepoint's "Discovery" DLP module scans assets attached to the network for sensitive data.
Features	• Scans the network for sensitive data, including data held in the cloud. • Categorizes data by level of sensitivity, including regulated data and intellectual property. • Utilizes a proprietary technology to fingerprint data, applying encryption and other controls based on the level of sensitivity. • Can identify sensitive data across both structured and unstructured data sets.

SOURCE: Forcepoint, 2019.

Table C.7
Forcepoint DLP Module: Network

	Description and Capabilities
Basic Description	Forcepoint's "Network" DLP module monitors the network for data transfers into and out of the network.
Features	• Combines AI, ML, and behavior analytics to identify attacks. • Identifies unauthorized data transfers out of the network and can apply appropriate controls. • Can combine user behavior analytics to aid security operators in determining whether a data transfer was malicious. • Behavior analytics can help identify and mitigate insider threat. • Captures event details from outside attacks. • Recognizes data within images.

SOURCE: Forcepoint, 2019.

References

Asen, Alex, Walter Bohmayr, Stefan Deutscher, Marcial Gonzalez, and David Mkrtchian, "Are You Spending Enough on Cybersecurity?" Boston Consulting Group, February 20, 2019.

Asti, Aric, *Cyber Defense Challenges from the Small and Medium-Sized Business Perspective,* Bethesda, Md.: SANS Institute, 2019. As of September 11, 2019:
https://www.sans.org/reading-room/whitepapers/hsoffice/cyber-defense-challenges-small-medium-sized-business-perspective-38160

Barth, Bradley, "Former NSA Director: Public and Private Sectors Must Unite Against Cyberattacks," *SC Magazine*, March 7, 2019. As of September 9, 2019:
https://www.scmagazine.com/home/security-news/former-nsa-director-public-and-private-sectors-must-unite-to-prevail-against-advanced-cyberattacks/

Bertuca, Tony, and Justin Doubleday, "Pentagon Reveals New Acquisition Initiatives to Block China," *Inside Defense*, August 26, 2019.

BlackBerry Cylance, "CylanceGUARD," webpage, 2019a. As of September 10, 2019:
https://www.cylance.com/en-us/platform/products/cylance-guard.html

BlackBerry Cylance, "CylanceOPTICS," webpage, 2019b. As of September 10, 2019:
https://www.cylance.com/en-us/platform/products/cylance-optics.html

BlackBerry Cylance, "CylancePROTECT," webpage, 2019c. As of September 10, 2019:
https://www.cylance.com/en-us/platform/products/index.html

BlackBerry Cylance, "World-Class Antivirus Protection," webpage, 2019d. As of September 10, 2019:
https://shop.cylance.com/us/features

Bourne, Vanson, *Underserved and Unprepared: The State of SMB Cyber Security in 2019*, Boston, Mass.: Continuum Managed Services, 2019.

Bromium, Inc., *The Hidden Costs of Detect-to-Protect Security*, Cupertino, Calif., 2018.

Brown, Michael, and Pavneet Singh, *China's Technology Transfer Strategy: How Chinese Investments in Emerging Technology Enable A Strategic Competitor to Access the Crown Jewels of U.S. Innovation*, U.S. Department of Defense, Defense Innovation Unit Experimental, January 2018.

"CISA Security Bill Passes Senate with Privacy Flaws Unfixed," *Wired*, October 27, 2015. As of September 10, 2019:
https://www.wired.com/2015/10/cisa-cybersecurity-information-sharing-act-passes-senate-vote-with-privacy-flaws/

CISA—*See* Public Law 114-113, Consolidated Appropriations Act, Division N: Cybersecurity Act of 2015, Title I: Cybersecurity Information Sharing Act (CISA), December 18, 2015.

Code of Federal Regulations, Title 32: National Defense, Subtitle A: Department of Defense, Chapter I: Office of the Secretary of Defense, Subchapter M: Miscellaneous, Part 236, Department of Defense (DOD)–Defense Industrial Base (DIB) Cybersecurity Activities.

Code of Federal Regulations, Title 32: National Defense, Subtitle B: Other Regulations Relating to National Defense, Chapter XX: Information Security Oversight Office, National Archives and Records Administration, Part 2002: Controlled Unclassified Information (CUI).

CrowdStrike, "Cyber Threat Intelligence Platform: Falcon X," webpage, 2019a. As of September 10, 2019:
https://www.crowdstrike.com/endpoint-security-products/falcon-x-threat-intelligence/

CrowdStrike, "Endpoint Detection & Response: Falcon Insight," webpage, 2019b. As of September 10, 2019:
https://www.crowdstrike.com/endpoint-security-products/falcon-insight-endpoin
t-detection-response/

CrowdStrike, "Endpoint USB Device Control: Falcon Device Control," webpage, 2019c. As of September 10, 2019:
https://www.crowdstrike.com/endpoint-security-products/falcon-endpoint-device-control/

CrowdStrike, "Fast, Effective Vulnerability Assessment: Falcon Spotlight," webpage, 2019d. As of January 17, 2020:
https://www.crowdstrike.com/endpoint-security-products/falcon-spotlight-vulnerabilit
y-management/

CrowdStrike, "Network Security Monitoring & IT Hygiene: Falcon Discover," webpage, 2019e. As of September 10, 2019:
https://www.crowdstrike.com/endpoint-security-products/falcon-discover-network-securit
y-monitoring/

CrowdStrike, "Next-Generation Antivirus (NGAV): Falcon Prevent," webpage, 2019f. As of September 10, 2019:
https://www.crowdstrike.com/endpoint-security-products/falcon-prevent-endpoint-antivirus/

CrowdStrike, "Proactive Threat Hunting: Falcon OverWatch," webpage, 2019g. As of September 10, 2019:
https://www.crowdstrike.com/endpoint-security-products/falcon-overwatch-threat-hunting/

D&B Hoovers, "First Rf Corporation," webpage, undated. As of November 15, 2019:
http://www.hoovers.com/company-information/cs/company-profile.first_rf_
corporation.60c2530b2a550019.html

D&B Hoovers, "MaXentric Technologies, LLC," webpage, undated, as of November 15, 2019:
http://www.hoovers.com/company-information/cs/company-profile.maxentric_technologies_
llc.367a0b864dc5b0da.html?aka_re=2

Defense Federal Acquisition Regulation Supplement 252.204-7000, "Disclosure of Information," October 2016. As of January 17, 2020:
https://www.acq.osd.mil/dpap/dars/dfars/html/current/252204.htm#252.204-7000

Defense Federal Acquisition Regulation Supplement 252.204-7012, "Safeguarding Covered Defense Information and Cyber Incident Reporting," October 2016. As of January 17, 2020:
https://www.acq.osd.mil/dpap/dars/dfars/html/current/252204.htm#252.204-7012

Department of Defense Instruction 5000.60, *Defense Industrial Base Assessments*, Washington, D.C.: U.S. Department of Defense, July 18, 2014.

Department of Defense Instruction 8010.01, *Department of Defense Information Network (DoDIN) Transport*, Washington, D.C.: U.S. Department of Defense, September 10, 2018.

DFARS—*See* Defense Federal Acquisition Regulation Supplement.

DIBNet, website, undated. As of January 17, 2020:
https://dibnet.dod.mil/portal/intranet/

Digital Guardian, *Endpoint Data Loss Prevention*, 2019. As of February 28, 2019:
https://info.digitalguardian.com/rs/768-OQW-145/images/DG-Endpoint-DLP.pdf

DODI—*See* Department of Defense Instruction.

Doubleday, Justin, "New Report Finds Defense Contractors Struggling with Cybersecurity Requirements," *Inside Defense,* May 21, 2019a.

Doubleday, Justin, "Pentagon to Require New Cybersecurity 'Certification' from Defense Contractors," *Inside Defense,* June 6, 2019b.

Doubleday, Justin, "New Pentagon Initiatives Seek to Overcome Entrenched Supply Chain Security Concerns," *Inside Defense,* September 9, 2019c.

Ernst & Young, "Is Cybersecurity About More Than Protection? EY Global Information Security Survey 2018–19," October 10, 2018. As of September 10, 2019:
https://www.ey.com/en_gl/advisory/global-information-security-survey-2018-2019

Executive Order 13556, *Controlled Unclassified Information*, Washington, D.C.: The White House, November 4, 2010.

Executive Order 13806, *Assessing and Strengthening the Manufacturing and Defense Industrial Base and Supply Chain Resiliency of the United States*, Washington, D.C.: The White House, July 21, 2017.

Federal Procurement Data System—Next Generation, "Top 100 Contractors Report," data sheets, 2006–2018. As of September 10, 2019:
https://www.fpds.gov/fpdsng_cms/index.php/en/reports/62-top-100-contractors-report

Federal Risk and Authorization Management Program, "Documents," webpage, undated. As of January 17, 2020:
https://fedramp.gov/documents/

Fidelis Cybersecurity, "Executive Summary," undated. As of September 10, 2019:
https://www.fidelissecurity.com/wp-content/uploads/2019/03/executivesummary.pdf

Fidelis Cybersecurity, "Fidelis Elevate™," data sheet, April 2019. As of September 11, 2019:
https://www.fidelissecurity.com/wp-content/uploads/2019/04/elevate.pdf

FireEye, "Email Security," webpage, 2019a. As of September 10, 2019:
https://www.fireeye.com/solutions/ex-email-security-products.html

FireEye, "Endpoint Security," webpage, 2019b. As of September 10, 2019:
https://www.fireeye.com/solutions/hx-endpoint-security-products.html

FireEye, "FireEye Security Suite," webpage, 2019c. As of September 10, 2019:
https://www.fireeye.com/solutions/security-suite.html

FireEye, "Helix Security Platform," webpage, 2019d. As of September 10, 2019:
https://www.fireeye.com/solutions/helix.html

FireEye, "Managed Defense," webpage, 2019e. As of September 10, 2019:
https://www.fireeye.com/solutions/managed-defense.html

FireEye, "Network Security and Forensics," webpage, 2019f. As of September 10, 2019:
https://www.fireeye.com/solutions/nx-network-security-products.html

FireEye, "Network Security and Forensics," webpage, 2019g. As of September 10, 2019:
https://www.fireeye.com/solutions/nx-network-security-products.html

Forcepoint, *Forcepoint Data Loss Prevention (DLP): Data Protection in a Zero-Perimeter World*, 2019. As of September 10, 2019:
https://www.forcepoint.com/sites/default/files/resources/files/brochure-dlp-en.pdf

Forrester Research, *2017 Tech Budget Benchmark*, Cambridge, Mass., March 28, 2017.

Friedberg Jr., Sydney J., "Top Official Admits F-35 Stealth Fighter Secrets Stolen," *Breaking Defense*, June 20, 2013.

Fruhlinger, Josh, "The State of IT Security, 2018," *CIO*, May 29, 2018.

Fung, Brian, "Apple and Dropbox Say They Don't Support a Key Cybersecurity Bill, Days Before a Crucial Vote," *Washington Post*, October 20, 2015.

Gartner, *IT Key Metrics Data 2017*, Stamford, Conn., December 12, 2016.

Gartner, *Magic Quadrant for Managed Security Services, Worldwide (ID: G00354867)*, May 2, 2019.

Gerberding, Katharina, "Cybersecurity Budgeting 101: How to Optimize Your Security Spend for Maximum ROI," *Hitachi Security Systems*, June 26, 2018.

Glassdoor, "First RF," webpage, undated-a. As of November 15, 2019:
https://www.glassdoor.com/Overview/Working-at-FIRST-RF-EI_IE940014.11,19.htm

Glassdoor, "MaXentric Technologies," webpage, undated-b. As of November 15, 2019:
https://www.glassdoor.com/Overview/Working-at-Maxentric-Technologies-EI_IE274980.11,33.htm

Glassdoor, "Cyber Security Salaries," webpage, 2019. As of September 10, 2019:
https://www.glassdoor.com/Salaries/cyber-security-salary-SRCH_KO0,14.htm

Goldman, Jeff, "Most Small to Mid-Sized Organizations Don't Use Multi-Factor Authentication," *eSecurity Planet*, August 16, 2018. As of September 7, 2019:
https://www.esecurityplanet.com/network-security/most-small-to-mid-sized-organizations-dont-use-multi-factor-authentication.html.

Gonzales, Daniel, Jeremy M. Kaplan, Evan Saltzman, Zev Winkelman, and Dulani Woods, "Cloud-Trust: A Security Assessment Model for Infrastructure as a Service (IaaS) Clouds," *IEEE Transactions on Cloud Computing*, Vol. 5, No. 3, July 2017, pp. 523–536. As of September 10, 2019:
https://doi.org/10.1109/TCC.2015.2415794

Grassi, Paul A., *Digital Identity Guidelines Authentication and Lifecycle Management*, NIST Special Publication 800-63B, Washington, D.C.: U.S. Department of Commerce, National Institute of Standards and Technology, 2017.

Guevara, Jamie, Eric Stegman, and Linda Hall, *Gartner IT Key Metrics Data 2012: IT Enterprise Summary Report*, Stamford, Conn.: Gartner, 2012.

Hauer, Barbara, *Data Leakage Prevention: A Position to State-of-the-Art Capabilities and Remaining Risk*, Institute of Systems Software, 2014.

Heidenreich, John, "The Privacy Issues Presented by the Cybersecurity Sharing Act," *North Dakota Law Review*, Vol. 91, 2015, pp. 395–410.

Inspector General, U.S. Department of Defense, *Audit of Protection of DoD Controlled Unclassified Information on Contractor-Owned Networks and Systems*, Washington, D.C., July 23, 2019.

Jackson Higgins, Kelly, "How Lockheed Martin's 'Kill Chain' Stopped SecurID Attack," *Dark Reading*, February 12, 2013. As of September 8, 2019,
https://www.darkreading.com/attacks-breaches/how-lockheed-martins-kill-chai
n-stopped/240148399.

Jaffer, Jamil N., "Carrots and Sticks in Cyberspace: Addressing Key Issues in the Cybersecurity Information Sharing Act of 2015," *South Carolina Law Review*, Vol. 67, 2016.

Jaycox, Mark, "EFF Opposes Cybersecurity Bill Added to Congressional End of Year Budget Package," Electronic Frontier Foundation, December 18, 2015. As of September 10, 2019:
https://www.eff.org/deeplinks/2015/12/statement-finalized-congressional-cybersecurity-bill

Joint Task Force Transformation Initiative, *Security and Privacy Controls for Federal Information Systems and Organizations*, NIST Special Publication 800-171, Revision 2, Washington, D.C.: U.S. Department of Commerce, National Institute of Standards and Technology, April 2013.

Katz, Justin, "Alarmed by Lack of Ongoing Research, Navy Cyber Group Seeks Defensive Tech from Industry," *Inside Defense*, July 15, 2019.

Kavanagh, Kelly, Toby Bussa, and Gorka Sadowski, "Magic Quadrant for Security Information and Event Management," Gartner, December 3, 2018. As of September 10, 2019:
https://www.gartner.com/doc/reprints?id=1-5VGLBIJ&ct=181129&st=sb.

Kesan, Jay P., and Carol M. Hayes, "Bugs in the Market: Creating A Legitimate, Transparent, and Vendor-Focused Market for Software Vulnerabilities," *Arizona Law Review*, Vol. 58, No. 3, 2016.

Lang, Juan, Alexei Czeskis, Dirk Balfanz, Marius Schilder, and Sampath Srinivas, "Security Keys: Practical Cryptographic Second Factors for the Modern Web," paper presented at Financial Cryptography 2016: Financial Cryptography and Data Security, Christ Church, Barbados, February 22–26, 2016.

Lord, Nate, "Data Protection: Data in Transit vs. Data at Rest," *Digital Guardian*, January 3, 2019. As of February 27, 2019:
https://digitalguardian.com/blog/data-protection-data-in-transit-vs-data-at-rest

Lubold, Gordon, and Dustin Volz, "Chinese Hackers Breach U.S. Navy Contractors," *Wall Street Journal*, December 14, 2018.

Lubold, Gordon, and Dustin Volz, "Navy, Industry Partners Are 'Under Cyber Siege' by Chinese Hackers, Review Asserts," *Wall Street Journal,* March 12, 2019.

Mansfield, Matt, "Cyber Security Statistics: Numbers Small Businesses Need to Know," *Small Business Trends*, blog, January 3, 2017. As of September 10, 2019:
https://smallbiztrends.com/2017/01/cyber-security-statistics-small-business.html.

Manta, "MaXentric Technologies, LLC," webpage, undated. As of November 15, 2019:
https://www.manta.com/c/mml6gy0/maxentric-technologies-llc

Multi-State Information Sharing and Analysis Center, "LockerGoga," security primer, SP2019-0611, March 2019.

National Archives, "CUI Category: Controlled Technical Information," webpage, undated. As of January 17, 2020:
https://www.archives.gov/cui/registry/category-detail/controlled-technical-info.html

National Center for the Middle Market, website, undated. As of January 17, 2020:
https://www.middlemarketcenter.org

Organisation for Economic Co-operation and Development, "Data: Enterprises by Business Size (Indicator)," webpage, undated. As of September 10, 2019:
https://data.oecd.org/entrepreneur/enterprises-by-business-size.htm

Organisation for Economic Co-operation and Development, "Glossary of Statistical Terms," webpage, last updated on December 2, 2005. As of September 10, 2019:
https://stats.oecd.org/glossary/detail.asp?ID=3123

Paletta, Damian, and Daisuke Wakabayashi, "Apple Piles On as Senate Debates Cyber Bill; Apple joins Twitter in Opposing Information Sharing Legislation," *Wall Street Journal*, October 21, 2015.

Pollard, Jeff, *Security Budgets 2019: The Year of Services Arrives*, Cambridge, Mass.: Forrester, December 17, 2018.

Public Law 107-347, Federal Information Security Management Act (FISMA), December 12, 2002.

Public Law 113-274, Cybersecurity Enhancement Act of 2014, December 18, 2014.

Public Law 114-113, Consolidated Appropriations Act, Division N: Cybersecurity Act of 2015, Title I: Cybersecurity Information Sharing Act (CISA), December 18, 2015.

Purdue CERIAS, "Facebook: Protecting A Billion Identities Without Losing (Much) Sleep," video, September 19, 2013. As of January 17, 2020:
https://www.youtube.com/watch?v=GDhpjbejuQo

PwC, *The Global State of Information Security Survey*, London, UK, March 10, 2017

Reed, Brian, and Deborah Kish, *Magic Quadrant for Enterprise Data Loss Prevention*, Gartner, February 16, 2017.

Ross, Ron, Kelley Dempsey, Patrick Viscuso, Mark Riddle, and Gary Guissanie, *Protecting Controlled Unclassified Information in Nonfederal Systems and Organizations*, NIST Special Publication 800-171, Revision 1, Washington, D.C.: U.S. Department of Commerce, National Institute of Standards and Technology, December 2016, updated June 2018.

Ross, Ron, Victoria Pillitteri, Kelley Dempsey, Mark Riddle, Gary Guissanie, *Protecting Controlled Unclassified Information in Nonfederal Systems and Organizations,* Draft NIST Special Publication 800-171, Revision 2, Washington, D.C.: U.S. Department of Commerce, National Institute of Standards and Technology, June 2019a.

Ross, Ron, Victoria Pillitteri, Gary Guissanie, Ryan Wagner, Richard Graubart, and Deborah Bodeau, *Protecting Controlled Unclassified Information in Nonfederal Systems and Organizations: Enhanced Security Requirements for Critical Programs and High Value Assets*, Draft NIST Special Publication 800-171B, June 2019b.

RSA, "RSA SecurID Hardware Tokens," webpage, undated. As of September 8, 2019:
https://www.rsa.com/en-us/products/rsa-securid-suite/rsa-securid-access/securid-hardware-tokens

Sera-Brynn, *Reality Check: Defense Industry's Implementation of NIST SP 800-171: Keen Insights from Certified Cybersecurity Assessors*, May 2019.

Small Business and Entrepreneurship Council, "Facts & Data on Small Business and Entrepreneurship," webpage, undated. As of September 11, 2019:
https://sbecouncil.org/about-us/facts-and-data/

Song, Yingo, Michael E. Locasto, Angelos Stavrou, Angelos D. Keromytis, and Salvatore J. Stolfo, "On the Infeasibility of Modeling Polymorphic Shellcode," *Proceedings of the 14th ACM Conference on Computer and Communications Security*, 2007. As of September 10, 2019:
http://web2.cs.columbia.edu/~angelos/Papers/2007/polymorph.pdf

"Theft of F-35 Design Data Is Helping U.S. Adversaries—Pentagon," Reuters, June 19, 2013.

Tracy, Abigail, "The Problems Experts and Privacy Advocates Have with the Senate's Cybersecurity Bill," *Forbes,* October 29, 2015. As of September 10, 2019:
https://www.forbes.com/sites/abigailtracy/2015/10/29/the-problems-experts-and-privacy-advocates-have-with-the-senates-cybersecurity-bill/#5944e5e06ebf

U.S. Census Bureau, "Number of Firms, Number of Establishments, Employment, Annual Payroll, and Estimated Receipts by Large Enterprise Receipt Sizes for the United States, NAICS Sectors: 2012," 2012 County Business Patterns and 2012 Economic Census, Washington D.C., 2012. As of September 11, 2019:
https://www.census.gov/data/tables/2012/econ/susb/2012-susb-annual.html

U.S. Code, Title 15: Commerce and Trade, Chapter 7: National Institute of Standards and Technology, Section 272: Establishment, functions, and activities.

U.S. Department of Defense, "DIBNet Portal," webpage, undated. As of September 11, 2019:
https://dibnet.dod.mil/portal/intranet/

U.S. Department of Defense, *Fiscal Year 2017: Annual Industrial Capabilities*, Washington, D.C., March 2018.

U.S. Department of Defense, *Assessing and Strengthening the Manufacturing and Defense Industrial Base and Supply Chain Resiliency of the United States: Report to President Donald J. Trump by the Interagency Task Force in Fulfillment of Executive Order 13806*, Washington, D.C., September 2018.

U.S. Department of Defense, "DOD Announces Fiscal Year 2019 University Research Funding Awards," April 3, 2019a. As of September 10, 2019:
https://www.defense.gov/Newsroom/Releases/Release/Article/1804268/dod-announces-fiscal-year-2019-university-research-funding-awards/

U.S. Department of Defense, "Cybersecurity Maturity Model Certification (CMMC): Draft CMMC Model Rev 0.4 Release and Request for Feedback," briefing, September 2019b.

U.S. Department of Homeland Security, "Alert (TA17-181A), Petya Ransomware," February 15, 2018. As of November 15, 2019:
https://www.us-cert.gov/ncas/alerts/TA17-181A

U.S. Department of Homeland Security and U.S. Department of Justice, *Guidance to Assist Non-Federal Entities to Share Cyber Threat Indicators and Defensive Measures with Federal Entities Under the Cybersecurity Information Sharing Act of 2015*, Washington, D.C.: June 15, 2016.

U.S. Department of Labor, Bureau of Labor Statistics, "Employer Costs for Employee Compensation—March 2019," June 18, 2019a.

U.S. Department of Labor, Bureau of Labor Statistics, "Information Security Analysts," September 4, 2019b.

U.S. Small Business Administration, "Table of Size Standards," webpage, August 19, 2019. As of September 12, 2019:
https://www.sba.gov/document/support--table-size-standards

Volz, Dustin, "Chinese Hackers Target Universities in Pursuit of Maritime Military Secrets," *Wall Street Journal*, March 5, 2019.

Weber, Rick, "Wireless Industry Warns of Costs, Other Concerns from NIST Cyber Standards for Defense Contractors," *Inside Defense,* August 23, 2019.

Youngren, Jan, "Hidden VPN Owners Unveiled: 99 VPN Products Run by Just 23 Companies," VPNpro, June 2, 2019. As of January 16, 2020:
https://vpnpro.com/blog/hidden-vpn-owners-unveiled-97-vpns-23-companies/

About the Authors

Daniel Gonzales is a senior scientist at the RAND Corporation. His research expertise centers on command and control, communications, and intelligence; electronic warfare; cybersecurity; digital forensics; and cyber supply chain risk management. He received his Ph.D. in theoretical physics from the Massachusetts Institute of Technology.

Sarah Harting is a senior defense analyst at the RAND Corporation. Her research focuses on U.S. defense strategy and doctrine, strategic planning, foreign policy, and national security issues. Harting received her M.A. in international security from Georgetown University.

Mary Kate Adgie is a research assistant at the RAND Corporation. Her research interests include national security, security cooperation, international economic development, and African affairs. Adgie received her B.A. in international relations and economics from the College of William and Mary.

Julia Brackup is a research assistant at the RAND Corporation. Her research focuses on U.S. defense strategy, foreign policy, and national security issues. Brackup received her B.A. in international affairs from Lafayette College.

Lindsey Polley is an assistant policy researcher at the RAND Corporation. Her research interests include cyber policy, cyber strategy, and cybersecurity; emergent and disruptive technology application to the military; electronic warfare; multidomain operations; and wargaming. Polley is pursuing a Ph.D. in policy analysis at the Pardee RAND Graduate School.

Karlyn Stanley is a lawyer and senior policy analyst at RAND Corporation. Her research focuses on data privacy, cyber crimes, and emerging technology issues. Stanley received her J.D. from the Columbus School of Law, Catholic University.